HYBRID NEURAL NETWORK AND EXPERT SYSTEMS

HYBRID NEURAL NETWORK AND EXPERT SYSTEMS

by

Larry R. Medsker

Department of Computer Science and Information Systems
The American University

Kluwer Academic Publishers
Boston/Dordrecht/London

Table of Contents

Preface

Three years ago, when I started presenting tutorials on the integration of neural networks and expert systems, I could uncover only enough work in this area to fill one page of references. Today we see rapidly growing interest and an order of magnitude more projects on hybrid systems that combine neural networks and expert systems. Working systems have been developed for demonstrating feasibility and some are actually in use in practical situations.

Several developments have stimulated these activities. Attention has grown in the research community, including a recognition among some AI researchers that combined approaches will be necessary if the remaining tough problems in AI are to be solved. The rapid developments in R&D for neural networks have produced many applications and development tools. Both the expert system and the neural network technologies are at stages in which useful hybrid systems are conveniently possible. The current opportunity is to develop more efficient and effective design and development techniques to enable widespread production of useful and practical systems. Work is also needed to clarify the range of appropriate problems. Progress in hybrid systems will advance more rapidly as we share ideas and experiences, allowing practical models and guidelines to emerge.

This book is a step toward summarizing the state of hybrid systems and disseminating information about working systems that illustrate the issues and opportunities in this field. While other intelligent technologies should be considered eventually, the current focus is on the two that have a peculiarly complementary nature: the logical, reasoning aspect of expert systems and the visual, pattern-oriented nature of neural networks. Together they represent the range of human intelligence that is difficult or impossible to simulate with either technology alone.

The first part of the book summarizes the concepts and principles of neural networks and expert systems that are relevant to the integration of the two. The status of research in hybrid systems is summarized, and initial models for integration are presented.

Next, five case studies are presented in detail so that the nature of the problems and the design and implementation processes are clear. The applications cover product design using the analysis of technical data, control systems using the analysis of detector and sensor data, and the monitor and control of chemical processes. An important thread is the need to process data, often in large quantities, for efficient use in decision making.

The final part of the book covers guidelines and methodologies for hybrid system development and surveys the current choices for development tools and environments. While systems are currently being developed and used, the next few years should see dramatic improvements in these areas, and the final chapter presents ideas about what we can expect.

This book contains the results of my literature survey and analysis of hybrid systems research and applications development. Some of the material also derives from my experience developing hybrid systems, and for that I gratefully acknowledge the contributions of my students and colleagues. In particular, the collaborations with Masud Cader, Dalila Benachenhou, and Harold Szu were essential. Students in my NSF-sponsored Research Experiences for Undergraduates (REU) programs and American University graduate students Fran Labate, Silvana Rubino, and Michael Bramante also made important contributions. Thanks go to Ron Sun at the University of Alabama for information on the AAAI-92 Workshop.

Special thanks go to the case study contributors for their pioneering work in hybrid systems applications and for their work in sharing the results in this book. I am particulary grateful to Ray Foss at DuPont; Alper Caglayan, Jim Mazzu, and Paul Gonsalves at Charles River Analytics, Inc.; Laveen Kanal and Srinivasan Raghavan at LNK Corp.; and Jim Hendler and Anne Wilson at the University of Maryland.

Finally, I appreciate the assistance of Alper Caglayan at Charles River Analytics, Inc. and Steve Ward and Marge Sherald at Ward Systems, Inc. in the use of NueX and NeuroShell, respectively. I also appreciate the use of some figures and text from my chapters in *Hybrid Architectures for Intelligent Systems* (CRC Press), *Expert Systems and Applied Artificial Intelligence* (Macmillan Publishing Co.), and *Design and Development of Expert Systems and Neural Networks* (Macmillan Publishing Co.).

Part I Fundamentals of Hybrid Systems

Chapter 1

Overview of Neural and Symbolic Systems

Expert system and neural network technologies have developed to the point that the advantages of each can be combined into more powerful systems. In some cases, neural computing systems are replacing expert systems and other AI solutions. In other applications, neural networks provide features not possible with conventional AI systems, and they could provide aspects of intelligent behavior that have thus far eluded the AI symbolic/logical approach.

Because of recent advances in neural network technology, hybrid intelligent systems can now address new problems. As these systems grow in number and importance, developers need frameworks for understanding the combination of neural networks and expert systems and will need models and guidelines for effective implementation. This chapter provides the fundamental concepts about neural networks and expert systems and gives examples to illustrate the potential of each of the technologies. On this basis, subsequent chapters address the combination of neural network and expert systems to produce hybrid systems.

1.1 Expert System Strengths and Limitations

Expert systems perform reasoning using previously-established rules for a well-defined and narrow domain. They combine knowledge bases of rules and domain-specific facts with information from clients or users about specific instances of problems in the knowledge domains of the expert systems.

An important advantage of expert systems is the ease with which knowledge bases can be modified. This is a result of the architecture (see Figure 1.1), which separates the knowledge base from the inference engine. As a result, changing the knowledge base does not require programming but can be done via word processing or an editor. This feature makes knowledge engineering accessible to a wider variety of analysts, end users, and experts.

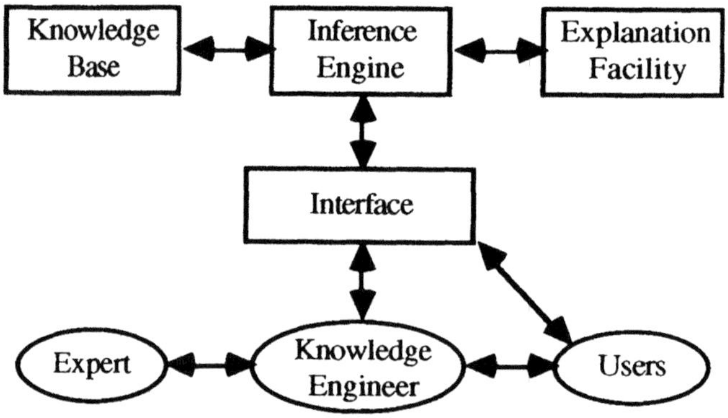

Figure 1.1. The architecture of expert systems

Ideally, reasoning can be explained, and most expert system development tools allow the creation of explanation systems to help the user understand questions being asked or conclusions. Thus, the system can function more like human experts who explain the reasoning processes behind their recommendations.

Expert systems are especially good for closed-system applications for which inputs are literal and precise, leading to logical outputs. For stable applications with well-defined rules, expert systems can be easily developed to provide good performance. Such systems can take advantage of the wealth of techniques developed in expert systems research to perform different types of reasoning using very efficient, systematic techniques. Thus, people are able to inspect and understand these systems because they have familiar structures similar to the logical frameworks humans use.

Another advantage of expert systems is the number and variety of commercial development systems that have become available over the last several years. These tools and associated techniques allow exploratory studies rapid prototyping for use in knowledge engineering. Prototypes are especially useful in gaining the interest and attention of experts so that the knowledge acquisition process is more productive and amicable.

A fundamental limitation of the expert system approach arises from the fact that experts do not always think in terms of rules. Thus, an expert system does not in these cases mimic the actual reasoning process of human experts. The computerized system is an attempt to produce performance that resembles human reasoning in some limited domain. The mechanism, however, may or may not resemble the actual biological or cognitive process. Thus, extensions of any expert system technique may not carry over into behavior that is similar to that of a human. Cognitive science, AI, and expert system research is still needed to produce a fundamental approach that models actual human reasoning.

In the meantime, a variety of approaches have been developed to address deficiencies in current expert system techniques, and many of these approaches show improvement in those limited areas. The major techniques of current interest are surveyed in the next section of this chapter. A comprehensive technique based on a good understanding of human reasoning is still needed.

A specific problem with expert systems is what is referred to as the knowledge acquisition bottleneck. While development tools have become very sophisticated and effective, expert systems still require extensive effort for eliciting knowledge from humans as well as from written material. Knowledge acquisition is still primarily a human-intensive activity requiring the usual system analysis abilities plus additional interviewing and interpersonal skills that are tailored to interacting with human experts. Furthermore, human experts may be too busy or otherwise difficult to deal with, so that a whole project may be threatened or delayed. Research and development efforts are focussing on understanding and automating the knowledge acquisition process to the extent possible. However, this activity remains a serious problem for the rapid and pervasive development of large expert systems.

Another difficulty with expert systems is in the area of large systems development. For real-world applications, the development process becomes difficult to manage. Working with experts and dealing with the complexity of large systems leads to prolonged, expensive development and delivery times. Furthermore, validation and verification of systems becomes difficult, if not impossible, as many lines of reasoning must be checked.

Other limitations involve fundamental uncertainties about the expert system approach. For example, more work is needed on how to represent commonplace knowledge, which humans deal with so well and so often. Also, expert systems do not automatically benefit from experience with their use and thus do not learn for failures or their use with novel examples.

While research and development are proceeding toward the improvement of expert system technology, some researchers question the underlying philosophy, at least in terms of the deficiencies described above. The alternate technologies discussed next address many of these concerns and may, along with solving certain practical problems, give insight into fundamental research questions that need to be answered before the goals for expert systems are fully realized.

1.2 Neural Computing

The state-of-the-art in neural computing is inspired by our current understanding of biological neural networks; however, after all the research in biology and psychology, important questions remain about how the brain and the mind work. Advances in computer technology allow the construction of interesting and useful artificial neural networks that borrow some features from the biological systems. Information processing with neural computers consists of analyzing patterns of activity, with learned information stored as weights between neurode connections (see Figure 1.2). A common characteristic is the ability of the system to classify streams of input data without the explicit knowledge of rules and to use arbitrary patterns of weights to represent the memory of categories. Together, the network of neurons can store information that can be recalled in order to interpret and classify future inputs to the network. Because knowledge is represented as numeric weights, the rules and reasoning process in neural networks are not readily explainable.

Neural networks have the potential to provide some of the human characteristics of problem solving that are difficult to simulate using the logical, analytical techniques of expert system and standard software technologies. For example, neural networks can analyze large quantities of data to establish patterns and characteristics in situations where rules are not known and can in many cases make sense of incomplete or noisy data. These capabilities have thus far proven too difficult for the traditional symbolic/logic approach.

Neural networks rely on training data to "program" the systems. Thus, neural network components can be useful for hybrid systems by using an appropriate training set that allows the system to learn and generalize for operation on future input data. Inputs exactly like training data are recognized and identified, while new data (or incomplete and noisy versions of the training data) can be put into the closest matches to patterns learned by the system.

Neural network components can be useful when rules are not known, either because the topic is too complex or no human expert is available. If training data can be generated, the system may be able to learn enough information to function as well as, or better than, an expert system. This approach also has the benefit of easy modification to a system by retraining with an updated data set, thus eliminating programming changes and rule reconstruction. The data-driven aspect of neural networks allows adjustment of changing environments and events. Another advantage of neural network components is the speed of operation after the network is trained, which will be enhanced dramatically as neural chips become readily available.

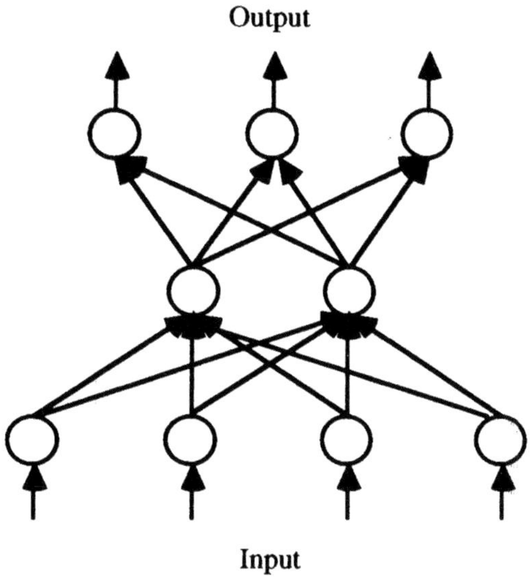

Figure 1.2. An example of nodes connected to form a neural network.

The immediate practical implications of neural computing are its emergence as an alternative or supplement to conventional computing systems and AI techniques. As an alternative, neural computing can offer the advantage of execution speed, once the network has been trained. The ability to learn from cases and train the system with data sets, rather than having to write programs, may be more cost effective and may be more convenient when changes become necessary. In applications where rules cannot be known, neural computers may be able to represent rules, in effect, as stored connection weights.

1.2.1 Neurons and Neurodes

The brain is composed of over 100 different kinds of special cells called neurons. The number of neurons in the brain is estimated to range from 50 to over 100 billion. The neurons are divided into interconnected groups called networks and provide specialized functions. Each group contains several thousand neurons that are highly interconnected with each other. Thus, the brain can be viewed as a collection of neural networks.

Thinking and intelligent behavior are believed to be controlled by the brain and the central nervous system. The ability to learn and react to changes in our environment requires intelligence. An example is the optical path in visual systems. External stimuli are transformed via cone cells and rod cells into signals that map features of the visual image into internal memory. Human intelligence is then used to understand the various visual features that are extracted and stored in memory.

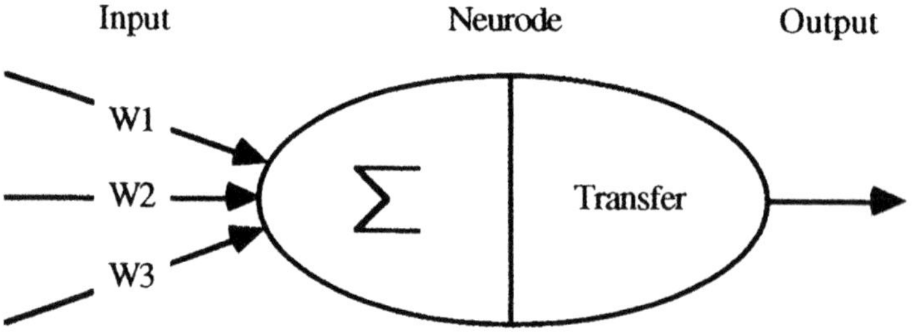

Figure 1.3. Model of a node in an artificial neural network.

An artificial neural network (ANN) is a model that emulates a biological neural network. The nodes in an ANN are based on the simplistic mathematical representation of what we think real neurons look like (see Figure 1.3). Today's neural computing uses a limited set of concepts from biological neural systems to implement software simulations of massively parallel processes involving processing elements (also called artificial neurons or neurodes) interconnected in a network architecture. The neurode is analogous to the biological neuron, receiving inputs that represent the electrical impulses that the dendrites of biological neurons receive from other neurons. The output of the neurode corresponds to signal sent out from a biological neuron over its axon. The axon of the biological neuron branches to the dendrites of other neurons and the impulses are transmitted over synapses. A synapse is able to increase or decrease its strength thus affecting the level of signal propagation and is said to cause excitation or inhibition of a subsequent neuron.

The history of artificial neural networks can be broken into an early period and decline in interest and then the recent period of intense research and development that has given rise to the neurocomputing field. In 1949, Donald Hebb wrote about early theories of neural learning that were soon applied to parallel distributed processing (PDP) networks. In the late 1950's, Frank Rosenblatt (Rosenblatt, 1962) developed the Perceptron model; however, in Minsky and Pappert, 1969 the single-layer perceptron was shown to be useless for a large class of problems. Recently, obstacles have been reduced by adding hidden layers and using learning algorithms such as back error propagation (backpropagation). However, right after Minsky's work, research interest and funding dropped off and the neural network field essentially lay dormant while the field of artificial intelligence flourished.

A few neural network researchers continued their work, however, and by the early 1980's the technological environment was ripe for a resurgence of interest. Hardware advances allowed fast processing, theoretical advances had been made in algorithms and computing techniques, and the prospects for parallel processing systems were good. Scientists and engineers such as Hopfield, Grossberg, Widrow, Anderson, and Kohonen discovered interesting ANN architectures and applications that demonstrated the potential of ANN technology. In the 1980's, rapid developments in research and development and the increasing appearance of applications has lead to the creation of neural network organizations, conferences, and journals. More recently, neural network development systems and tools are commercially available. As with expert systems, the availability of a convenient development method is allowing the spread of neurocomputing and is putting neurocomputing on the road to being in the standard repertoire of systems developers.

An important function of the artificial neuron is the evaluation of its inputs and the production of an output response. A weighted sum of the inputs from the simulated dendrites is evaluated to determine the level of the output on the simulated axon. Most artificial systems use threshold values and a common activation function is the sigmoid function, f, that can squash the total input summation to a bounded output value.

This model of the neuron, or basic perceptron, requires a learning algorithm for deriving the correct weights that correctly represent the knowledge to be stored. A fundamental concept in that regard is Hebbian learning, based on Donald Hebb's work (see Hebb, 1949) on biological systems, which postulates that active connections should be reinforced. This means that the strengths (weights) of interconnections increase if the prior node repeatedly stimulates the subsequent node to generate an output signal. In some algorithms, the weights of connections may also be decreased if they are not involved in stimulating a node, and negative weights can also be used to represent inhibiting actions.

For more complex neural computing applications, neurodes are combined together in various architectures useful for information processing (see Figure 1.4). A common arrangement has layers of neurodes with forward connections to every neurode except those in the same or prior layer. Useful applications require multiple (hidden) layers between the input and output neurodes and a correspondingly large number of connections.

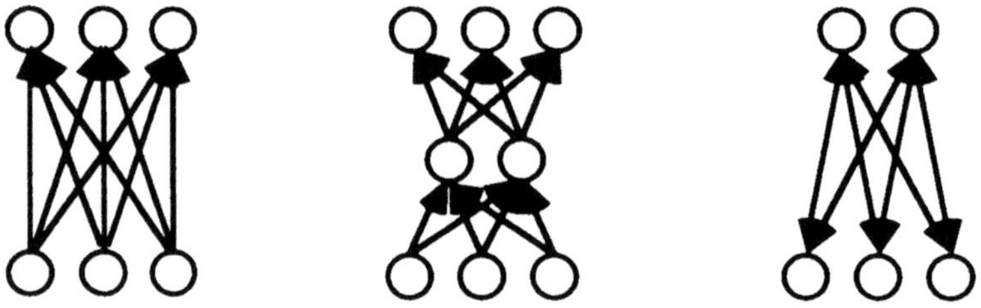

Associative Memory | Hidden Layer | Double Layer

Figure 1.4. Examples of neural network architectures.

Information processing with neural computers consists of analyzing patterns of data, with learned information stored as neurode connection weights. A common characteristic is the ability of the system to classify streams of input data without the explicit knowledge of rules and to use arbitrary patterns of weights to represent the memory of categories. During the learning stages, the interconnection weights change in response to training data presented to the system. In contrast, during recall the weights are fixed at the trained values.

Although most applications use software simulations, neural computing will eventually use parallel networks of simple processors that use the strengths of the interconnections to represent memory. Each processor will compute node outputs from the weights and input signals from other processors. Together, the network of neurons can store information that can be recalled in order to interpret and classify future inputs to the network.

The state-of-the-art in neural computing is inspired by our current understanding of biological neural networks. However, despite the extensive research in neurobiology and psychology, important questions remain about how the brain and the mind work. This is just one reason why neural computing models are not very close to the current understanding of biological systems.

Research and development in the area of ANN are producing interesting and useful systems that borrow some features from the biological systems, even though we are far from having an artificial brain-like machine. The field of neural computing is in its infancy with much research and development required in order to mimic the brain and mind. However, many useful techniques inspired by the biological systems have already been developed and are finding use in real-world applications.

1.2.2. Neurocomputing Fundamentals

The key terms needed to understand artificial neural networks are as follows:

a. The neurode. An ANN is composed of basic units called artificial neurons, or neurodes, that are the processing elements (PEs) in a network. Each of the neurodes receives input data, processes the input, and delivers a single output. This process is shown in Figure 3. The input can be raw data or output of other PEs. The output can be the final product or it can be an input to another neurode.

b. A network. An ANN is composed of a collection of interconnected neurons that are often grouped in layers; however, in general, no specific architecture should be assumed. The various possibilities of neural network topologies are the subject of research and development efforts. In terms of layered architectures, two basic structures are two layers, input and output, and three layers input, intermediate (hidden) and output. An input layer receives data from the outside world and sends signals to subsequent layers. The outside layer "interprets" signals from the previous layer to produce a result to the outside world as the network's "understanding" of the input data.

c. Inputs. Each input corresponds to a single attribute of a pattern or other data in the external world. The network can be designed to accept set of input values that are binary-valued or continuous-valued. For example, if the problem is to decide whether or not to approve a loan, an attribute can be an income level, age, etc. Note that in neurocomputing, we can only process numbers. Therefore, if a problem involves qualitative attributes or graphics, the information must be pre-processed to a numerical equivalence before it can be interpreted by the ANN.

Examples of inputs to neural networks are pixel values of characters and other graphics, digitized images and voice patterns, digitized signals from monitoring equipment, and coded data from loan applications. In all cases, an important initial step is to design a suitable coding system so that the data can be presented to the neural network, commonly as sets of 1's and 0's. For example, a 6x8-pixel character would be a 48-bit vector input to the network.

d. Outputs. The output of the network is the solution to the problem. For example, in the loan application case it may be "yes" or "no." The ANN, again, will assign numerical values (e.g. +1 means yes; zero means no). The purpose of the network is to compute the value of the output. In the unsupervised type of ANN, the initial output of the network is usually incorrect and the network must be adjusted or tuned until it gives the proper output.

e. Hidden Layers. In multi-layered architectures, the inner ("hidden") layers do not directly interact with the outside world, but add a degree of complexity to allow the ANN to operate on more interesting problems. The hidden layer adds an internal representation of the problem that provides the network the ability to deal robustly with inherently nonlinear and complex problems.

f. Weights. The weights in an ANN express the relative strengths (or mathematical value) of the various connections that transfer data from layer to layer. In other words, the weights express the relative importance of each input to a PE. Weights are crucial to ANN because they are the means by which we repeatedly adjust the network to produce desired outputs and thereby allow the network to "learn." The objective in training a neural network is to find a set of weights that will correctly interpret all the sets of input values that are of interest for a particular problem. Such a set of weights is possible if the number of neurodes, the architecture, and the corresponding number of weights form a sufficiently complex system to provide just enough parameters to adjust (or "tune") to produce all the desired outputs.

g. Summation function. The summation function finds the weighted average of all the input element. A simple summation function will multiply each input value (X_j) by its weight ($W_{i,j}$) and total them together for a weighted sum S. The formula for N input elements is $S = \Sigma w_i * x_i$. The neurodes in a neural network thus have very simple processing requirements, needing mainly to monitor the incoming signals from other neurodes, compute the weighted sums, and determine a corresponding signal to send to other neurodes.

h. Transformation function. The summation function computes the internal stimulation or activation level of the neuron. Based on this level the neuron may or may not produce an output. The relationship between the internal activation level may be linear or may be nonlinear. Such relationships are expressed by a transformation function. The sigmoidal function, which is commonly and effectively used, is discussed below. The selection of the specific function, as well as that of the transformation function, is one of the variables in choosing a network architecture and learning paradigm. Although many different functions are possible, a very useful and popular nonlinear transfer function is the sigmoid (or logical activation) function.

The collective action of a neural network is like a committee or other group making a decision. Individuals interact and effect each other in the process of arriving at a group decision. The global average or consensus of the group is more significant than an individual opinion and can even remain the same if some individuals drop out of the group. Also, a group can have different mechanisms for arriving at the collective decision.

i. Learning (see Figure 1.5). The sets of weight values for a given neural network represent different states of its memory or understanding of the possible inputs. In supervised networks, training involves the adjustment of the weights to produce the almost correct outputs. Thus, the network "learns" how to respond to patterns of data presented to it. In other types of ANN's, the networks self-organize and learn categories of input data.

1.2.3 Network Architectures

Many different neural network models and implementations are being developed and studied today. Three representative architectures (with appropriate learning paradigms) are shown in Figure 4 and are discussed next.

a. Associative memory systems. These systems correlate input data with information stored in the memory. Information can be recalled from incomplete or noisy input and the performance degrades only slowly as neurons fail. Associative memory systems can detect similarities between new input and stored patterns. Most neural network architectures can be used as associative memories, and a prime example is the Hopfield network (Hopfield, 1988).

b. Multiple-layered systems. Associative memory systems can have one or more intermediate (hidden) layers. An example of a simple network is shown in Figure 1.4. The most common learning algorithm for this architecture is back-error propagation, which is a kind of credit-blame approach to correcting and reinforcing the network as it adjusts to the training data presented to it. Another type of supervised learning, competitive filter associative memory, has capabilities for learning by changing its weights in recognition of categories of input data without being provided examples by an external trainer. A leading example of such a self-organizing system for a fixed number of classes in the inputs is the Kohonen network.

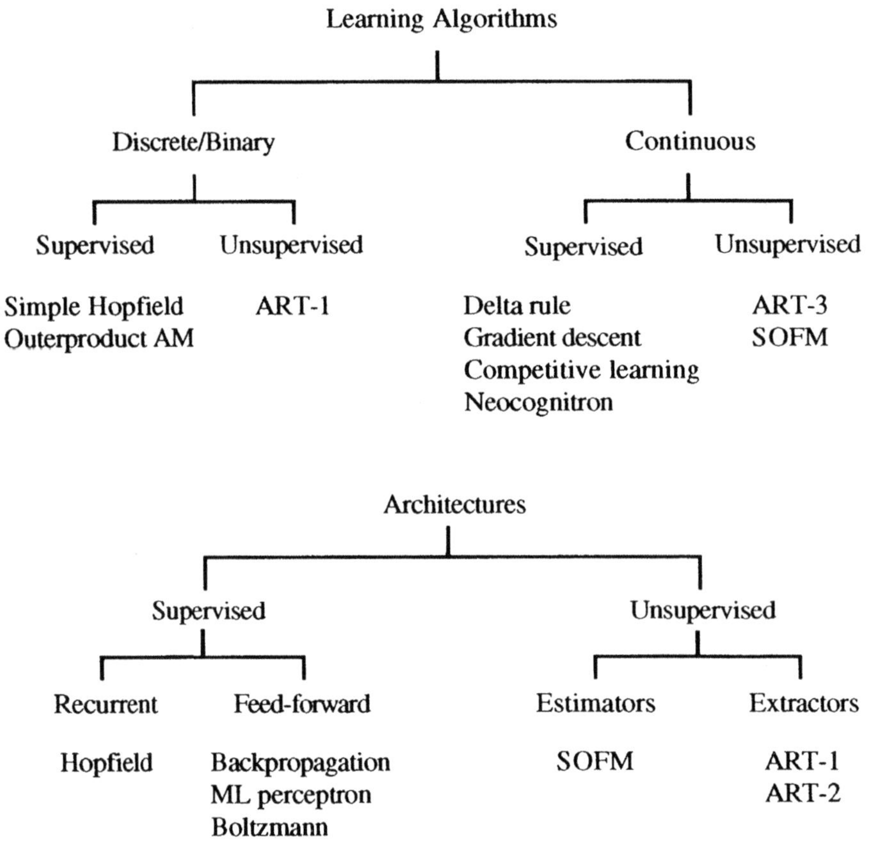

Figure 1.5. Taxonomy of artificial neural network architectures and learning algorithms

c. <u>A double layer structure</u>. A double layer structure, exemplified by The Adaptive Resonance Theory (ART) approach (Carpenter and Grossberg, 1987), does not require the knowledge of a precise number of classes in the training data but uses feed forward and backward to adjust parameters as data is analyzed to establish arbitrary numbers of categories that represent the data presented to the system. Parameters can be adjusted to tune the sensitivity of the system and produce meaningful categories.

1.2.4 Learning Paradigms

An important consideration in ANN is the appropriate use of algorithms for learning. ANN's have been designed for different type of learning:

- hetero-association - mapping one set of data to another. Produces an output not generally the same as the form of the input pattern. Used, e.g., in stock market prediction applications.
- auto-association - storing patterns for error tolerance. Reproduces an output pattern similar to or exactly the same as the input pattern. Used in optical character recognition systems.
- regularity detection - looking for useful features in data (feature extraction). E.g., in sonar signal identification systems.
- reinforcement learning - acting on feedback. Supervised form of learning where the teacher is more of a critic than an instructor. Used in controllers in ultrasonic spaceplane.

Two basic approaches to learning in an ANN exist: supervised and unsupervised learning:

Supervised learning. In this approach, we use a set of inputs for which the appropriate outputs are known. In one type, the difference between the desired and actual output is used to calculate corrections to the weights of the neural network (learning with a teacher). A variation on that approach simply acknowledges for each input trial whether or not the output is correct as the network adjusts weights in an attempt to achieve correct results (reinforcement learning).

Unsupervised learning. In this case, the neural network self-organizes to produce categories into which a series of inputs fall. No knowledge is supplied about what classifications are correct, and those that the network derives may or may not be meaningful to the person training the network. However, the number of categories into which the network classifies the inputs can be controlled by varying certain parameters in the model. In any case, a human must examine the final categories to assign meaning and determine the usefulness of the results. Examples of this type of learning are the Adaptive Resonance Theory (Carpenter and Grossberg, 1987) and Kohonen Self-Organizing Feature Maps (Kohonen, 1988).

Although many supervised learning examples exist, other important cases, such as the exclusive OR, cannot be handled with a simple neural network. Patterns must be linearly separable -- i.e., in the x-y plot of pattern space, it must be possible to draw a straight line that divides the clusters of input-output points that belong to the desired categories. In the example above, the input-output pairs (0,1), (1,0), and (1,1) are linearly separable from (0,0). Although the requirement of a linearly separable input pattern space caused initial disillusionment with neural networks, recent models such as back propagation in multi-layer networks have greatly broadened the range of problems that can be addressed.

Back-error propagation (Werbos, 1993), a popular technique that is relatively easy to implement, requires training data for providing the network with "experience" before using it for processing other data. Externally-provided correct patterns are compared with the neural network output during training, and feedback is used to adjust the weights until all training patterns are correctly categorized by the network. In some cases, a disadvantage of this approach is prohibitively large training times.

For any output neuron, j, the error delta = $(Zj-Yj)*f'$, where Z and Y are the desired and actual outputs, respectively, and f' is the slope of a sigmoid function evaluated at the jth neuron. If f is chosen to be the logistic function, then $f' = df/dx = f(1-f)$, where $f(x) = [1+\exp(-x)]^{-1}$ and x is proportional to the sum of the weighted inputs to the of the jth neuron. A more complicated expression can be derived to work backwards from the output neurons through the inner layers to calculate the corrections to their associated weights.

The procedure for executing the learning algorithm is as follows: initialize weights and other parameters to random values, read in the input vector and desired output, calculate actual output via the calculations forward through the layers, change the weights by calculating errors backward from the output layer through the hidden layers. This procedure is repeated for all the input vectors until the desired and actual outputs agree within some predetermined tolerance.

1.2.5 Advantages and Limitations of Neural Networks

Although ANN's have exciting possibilities, they are of course not without their limitations. Traditional artificial intelligence approaches have in their favor the more transparent mechanisms often expressed in terms, such as logic operations and rule-based representations, that are meaningful to us in our everyday lives. On the other hand, an ANN does not use structured knowledge with symbols used by humans to express reasoning processes. Furthermore, ANN's have so far been used for classification problems and, although quite effective at that type of task, need to be expanded to other types of intelligent activities.

An ANN's weights, even though quite effective, are just a set of numbers that in most cases has no obvious meaning to humans. Thus, an ANN is a "black box" solution to problems, and an explanation system cannot be constructed to justify a given result. As noted before, another limitation can be excessive training times, for example in ANN's using back-error propagation.

Neurocomputing is a relatively new field, and more research and development efforts in the future will surely minimize the limitations and find further strengths of this approach. Anticipated advances are in the improvement of fault tolerance aspects of ANN's, allowing them to be effective as individual neurodes fail or have incorrect input. The exciting prospects of self-organizing networks will be exploited to produce systems that learn on their own how to categorize input data. Future systems will improve in the areas of generalization and abstraction, being able to go beyond the training data to interpret patterns not explicitly seen before. Finally, the collaboration between scientists in neurocomputing and neurobiology should lead to advances in each field as computers mimic what we understand of human thinking and as neuroscientists learn from computer simulations of the theories of human cognition.

1.3 Using Neural Networks in Symbolic Processing Applications

In several areas of traditional AI applications, neural networks can be more effective substitutes and in other cases can work synergistically. The following sections present a brief survey of current applications of neural networks areas previously addressed only with AI systems.

Neural Networks in Natural Language Processing. Goals of R&D in this area include the ability to find correct interpretations of works in surrounding text, from written text or from spoken words and the translation of one language to another. For speech recognition, neural networks store, via training, information on speech parts for later rapid matching to input patterns. In one application, a front-end network recognizes short, phoneme-like fragments of speech and another network constructs words from combinations of the fragments. Another component clarifies ambiguities between words with similar sounds.

Neural Networks in Robotics, Vision, and Signal Processing. Aspects of this area include processing and understanding of sensor data, coordination between visual perception and mechanical action, sensing the context of the local environment, and the ability to learn and adapt. In the vision component of robotic systems, neural networks can use the associative memory feature to learn to interpret visual data such as partially obscured faces or objects and choose a close match with an image in memory. The systems take incoming visual data and extract features as sub-tasks of larger systems that use the feature extraction information.

Much of the work in robotic learning is in the simulation stage; however, the research results are encouraging. Recent work aims for systems that can interpolate learned data to create smoother motions and vary speeds as needed in specific situations. In the area of robot control, two aspects under intense study are path and trajectory planning and nonlinear control of motors and gears. Neural network systems are being developed for obstacle detection and adaptive response and for coordination of robot arms with input from cameras.

Handwriting Recognition. Another use of neural networks is the recognition of handwritten characters. Automatic verification of signatures on checks and other documents could save processing costs and reduce losses due to unauthorized transactions. Machine reading of forms filled out in handwriting advantageous because people do not always have convenient access to typewriters and they tend to make fewer mistakes if forms are entered by handwriting. The range of characters that can be analyzed includes graphics symbols and Japanese and Chinese characters. Some systems read words directly from paper documents so that a person does not have to type, while other systems have a person use a light pen or write on a sensitive panel with a stylus.

Applications for Decision Support Systems. An important area of information systems deals with tools and techniques that aid decision makers, especially at the middle and top levels of management. The domain of applicability includes dynamic, open systems subject to considerable uncertainty and risk. Problems tend not to be well structured, and exact solutions and data requirements are difficult to anticipate. Traditional decision support systems provide immediate access and flexible analysis of data, including access to a variety of databases and models as required. The components of such systems include database management systems, model base management systems, and facilities for generating dialogue interfaces. Tools need to enable decision makers to estimate consequences of proposed actions and to model situations for finding optimal solutions.

Neural networks can provide capabilities not available in decision support systems or expert systems -- specifically, the ability to adapt to new situations typical of open systems and to generalize from experience and interpolate from learned facts to recognize similar situations. A natural application of neural networks is the processing of large data sets to identify patterns and features that require further attention and may reveal the need for decisions. Neural networks could be components of database mining systems that run in the background or overnight to look for problems or interesting correlations in a database that may be of interest to a managerial decision maker. A goal for intelligent database systems is to handle information and decision making in a way more similar to humans, and the neural network components may be crucial for finding patterns in data, finding approximate matches and best guess estimates, and facilitating inexact queries.

1.4 Neural Network and Expert System Comparisons

Beyond its role as an alternative, neural computing can be combined with conventional software to produce more powerful hybrid systems. Such integrated systems could use database, expert system, neural network, and other technologies to produce the best solutions to complex problems. Thus, intelligent systems could eventually mimic human decision making under uncertainty and where information is incomplete or contains mistakes. A goal is to produce systems that include components that exhibit mind-like behavior in order to handle information as flexibly and powerfully as humans do.

Expert systems and artificial neural networks have unique and sometimes complementary features. From functional and applications standpoints, each approach can be equally feasible, although in some cases one may have an overall advantage over the other. In principle, expert systems represent a logical, symbolic approach while neural networks use numeric and associative processing to mimic models of biological systems. Features of each approach are summarized as follows:

Expert Systems	Neural Networks
Symbolic	Numeric
Logical	Associative
Mechanical	Biological
Sequential	Parallel
Closed	Self-organizing

Neural networks rely on training data to "program" the systems. Thus, neural network components can be useful for hybrid systems by using an appropriate training set that allows the system to learn and generalize for operation on future input data. Inputs exactly like training data are recognized and identified, while new data (or incomplete and noisy versions of the training data) can be put into the closest matches to patterns learned by the system.

Neural network components can be useful when rules are not known, either because the topic is too complex or no human expert is available. If training data can be generated, the system may be able to learn enough information to function as well as, or better than, an expert system. This approach also has the benefit of easy modification to a system by retraining with an updated data set, thus eliminating programming changes and rule reconstruction. The data-driven aspect of neural networks allows adjustment of changing environments and events. Another advantage of neural network components is the speed of operation after the network is trained, which will be enhanced dramatically as neural chips become readily available.

The two technologies in many ways represent complementary approaches, and neural network components can be the best solutions for some of the problems that have proven difficult for expert system developers and allow system developers to address problems not amenable to either approach alone. Developers need models and guidelines for making good use of the new opportunities presented by the synergism of neural networks and expert systems, and they need to know when to choose between each technology and how to implement systems that combine the two effectively. However, the integration of these and other intelligent systems components with conventional technologies promises to be an important area for research and development in the 1990's.

The similarities between expert system and neural network development and their complementary nature make hybrid systems a very promising area for research and development. The integration of these and possibly other intelligent components with conventional computing technologies promises to be an important area for both research and development and for applications in the 1990's.

Chapter 2

Research in Hybrid Neural and Symbolic Systems

The emerging need for hybrid connectionist and symbolic systems is currently motivating important research and development work. In terms of the individual technologies, connectionist systems can represent the perceptual aspect of intelligence while symbolic systems provide the cognitive features. In addition to the practical aspect, systems that perform both perceptual as well as cognitive types of tasks are interesting also for expanding the AI effort to better understand human cognition and model it in computer systems.

This chapter gives an overview of the areas of research and development on fundamentals of hybrid systems, including theoretical issue and integration techniques. This is a broader view of hybrid intelligence, beyond the integration of neural networks and expert systems. For more details, summaries of recent research can be found in AAAI, 1990; Frisch and Scherl, 1991; Frisch and Cohn, 1991; Kandel and Langholz, 1992; Soucek, 1992; Sun, Bookman, and Shekhar, 1992, Sun and Bookman, 1993, and Wilson and Hendler, 1993.

2.1 Hybrid Reasoning

As discussed in Hendler, 1989, four approaches to systems with cognitive and perceptual features are (see Figure 2.1)

1) develop AI systems that can handle well the perceptual, pattern recognition, and image and signal processing tasks;
2) discover a method for connectionist systems to handle high-level symbol-

processing tasks;
3) develop another paradigm other than the symbolic and connectionist approaches;
4) produce hybrid systems that exploit the strengths of the current generation of connectionist and symbolic technologies.

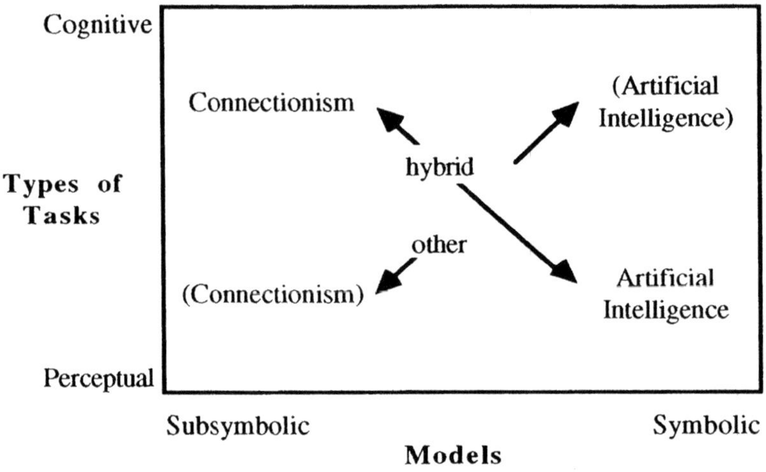

Figure 2.1 Four approaches to systems with cognitive and perceptual features.

Although the fourth approach has its own research and development challenges, recent advances in the design and use of hybrid systems are encouraging. Several successful applications have been developed, and hybrid system development tools and environments are becoming commercially available. The hybrid approach also makes sense in terms of modelling human intelligence, which has quite diverse aspects and capabilities (e.g., in tasks involving planning or language). Computer simulations of intelligence should take advantage of any appropriate technologies in order to reflect that diversity.

2.2 Hybrid Systems Research Areas

For the purpose of this chapter, the term hybrid refers to systems that consist of two or more integrated subsystems, each of which may have a different representation language and inference technique. The subsystems are assumed to be tied together semantically and influence each other. The coupling can be to such a degree that any given operation may be difficult to classify as symbolic or subsymbolic. On the other hand, loose couplings join distinct components and regulate the flow of information between them.

Hybrid systems research areas include translation of data from neural networks for use by expert systems, identifying appropriate problems for hybrid system, communication protocols for symbolic/subsymbolic interfaces, and multiple processor architectures (see, e.g., AAAI, 1991; Frisch and Cohn, 1991; Kandel and Langholz, 1992; and Soucek, 1991). The intelligent components can use techniques such as genetic algorithms and fuzzy logic, as well as neural networks and expert systems.

As shown in Figure 2.2, various ways are possible to combine reasoners and representations (for more detail see Frisch and Cohn, 1991). Multiple representations may be used to provide redundancy for some knowledge bases, while multiple representations may pertain to different types of knowledge. Reasoners may operate on the same representations or may each be used for particular representations. Hybrid reasoners can be general purpose, and therefore relatively slow, or specialized and fast when invoked be a primary reasoner needing a particular task performed.

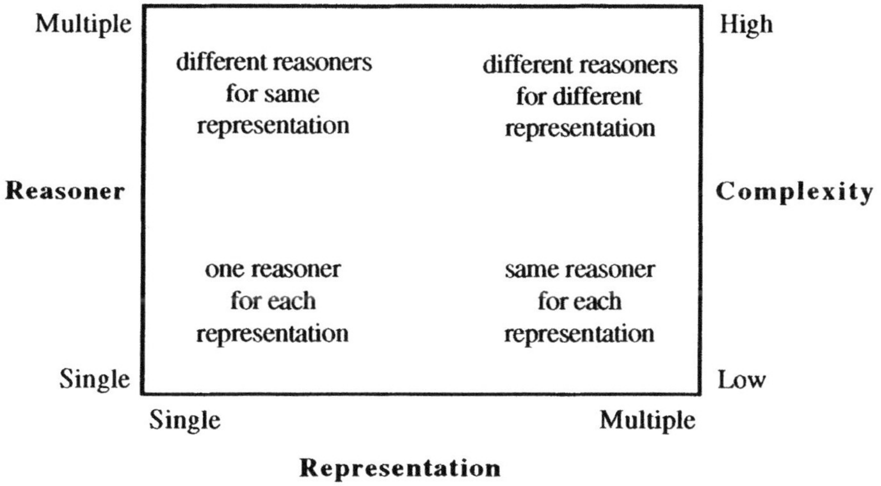

Figure 2.2 The various combinations of reasoners and representations.

2.2.1 Language Understanding

Research in this area seeks to add connectionist components to improve the effectiveness of traditional AI techniques for language recognition and understanding. As part of this work, fundamental aspects of the linkage between symbolic and subsymbolic components are investigated.

Some systems take as input sentences or phrases and transform them into activation patterns for input nodes of connectionist systems. The output can be associated with most likely interpretations of the input based on previous learning by the network. The output may also activate a semantic network to link the subsymbolic component to higher level symbols.

Some of this research explores ways to create networks automatically and dynamically as a function of the input sentences or phrases, and other approaches use manually created networks. For each input, a unique network may be created or the selection from a set of pretrained networks may be made.

Selected research in this area of language understanding can be found in the following references:

Topic	Reference
Two-tier framework for text comprehension	Bookman, 1992
Deductive and inductive learning in a rule-based connectionist parser	Kwasny and Faisal, 1991
Sentence understanding	Lange, 1991
Control of activation and inferencing	Lange, 1992
Sentence analysis	Lehnert, 1991
Parallel distributed semantic networks	Sumida, 1992
Phrase understanding	Wermter and Lehnert, 1989 Wermter, 1992

2.2.2 Representation Issues

This areas includes frameworks and architectures for integrating intelligent components. Design principles are investigated and structures and process are developed and tested to gain insight into the problems and solutions for providing for processing, learning, and controlling tasks in hybrid systems. Work in this area addresses distributed control structures that coordinate, switch between, choose, etc. the actions of independent subsystems. In the area of performance, better symbolic to subsymbolic mappings can allow the use of subsymbolic systems that were determined *a priori* and reduce the amount of learning during operational use of the hybrid system.

Topic	Reference
Linking domain knowledge to neural network structure	Fu, 1992
Representing expert knowledge in neural networks	Knaus, 1992
Learning in expert networks	Lacher, Hruska, and Kuncicky, 1991
Integration of learning into expert systems	Romaniuk and Hall, 1991 Hall and Romaniuk, 1992
Marker passing over microfeatures	Hendler, 1989
Deductive and inductive learning in a rule-based connectionist parser	Kwasny and Faisal, 1991
Framework for generalized connectionist networks	Honovar, 1990
Models for integration	Medsker and Bailey, 1992
Multilevel processing control	Lee and Nguyen, 1991
Distributed and local representations for complex knowledge structures	Samad, 1992
Superpositional representations	Sharkey, 1992

Tree matching with recursive distributed representations	Stolke and Wu, 1992
Hierarchical structure for expert system, neural network, decision tables, and barrels	Villa and Reilly, 1992
Fuzzy connectionist expert system	Machado and Rocha, 1992
Fuzzy hybrid systems	Posey, Kandel, and Langholz, 1992

2.2.3 Logic and Reasoning

In this area, researchers look at the ways to implement reasoning using connectionist models. An important issue is whether connectionist components can deal with aspects of reasoning that are difficult or inefficient to do with symbolic techniques. As discussed in Frisch and Cohn, 1991, one approach is to have connectionist nodes correspond to components of the symbolic system.

Hybrid system research also addresses ways to increase the efficiency of deduction by reducing search space in logic systems. Work also seeks to enhance the understanding of reasoning by analyzing the performance of experimental hybrid systems for problem solving and other logic operations.

Topic	Reference
RHET hybrid reasoning system	Allen and Miller, 1988
Belief representation	Barnden, 1992
KRYPTON hybrid reasoning system	Brachman, Gilbert, and Levesque, 1985
Symbolic/subsymbolic problem solver	Callan, 1990
Approximate reasoning in knowledge-based systems	Cohen and Hudson, 1992
Many-sorted logics	Cohn, 1989
Equivalence of rule-based expert systems and neural networks	Kuncicky, Hruska, and Lacher, 1991

Hierarchical network for reasoning	Lacher, 1992
Unrestricted first-order logic formulas	Pinkas, 1992
Theory resolution	Stickel, 1985
Connectionist model of rule-based reasoning	Sun and Waltz, 1991 Sun, 1992
Rule-combined neural network for design optimization	Yang, 1992

2.2.4 Coupling Mechanisms

This areas includes various interface issues, including communication between subsystems. Also in this category is the practical aspect of linking symbols with connectionist system components. Research into systems design aims both at useful tools for efficient and effective development of systems, but also fundamental principles that can be gained from understanding how hybrid systems work.

Topic	Reference
Linking rules to neural network nodes	Cohen and Hudson, 1992
Neural network learning and expert systems	Gallant, 1993
Interactions for control systems	Gomi and Nakamura, 1990
SCRuFFy - Integration of perceptual and symbolic reasoning	Hendler and DIckens, 1991
Equivalence of rule-based expert systems and neural networks	Kuncicky, Hruska, and Lacher, 1991
Using neural networks to establish rules	Towell and Shavlik, 1992
Linking symbolic and subsymbolic computing	Wilson and Hendler, 1993

The goals of hybrid system research include techniques to increase the efficiency, expressive power and reasoning power of intelligent systems. Some of the work on efficiency makes use of specialized reasoners strategically called by control or supervisor modules that decide which reasoners to use at different times.

A general goal is to create hybrid systems that are significantly more powerful, with less development effort, than systems that use symbolic or subsymbolic methods alone. Hybrid systems are able to address some practical problems that have thus far alluded traditional AI approaches. From the perspective of fundamentals, hybrid systems may also give further insight into cognitive mechanisms and models.

2.3 AAAI-92 Workshop

Much of the current research in hybrid systems was discussed at the 1992 AAAI Workshop on Integrating Neural and Symbolic Processes (see Sun, Bookman, and Shekhar, 1992 and Sun and Bookman, 1993). The focus of this workshop was the architectural approaches that relate to cognitive modeling. Issues include the types of appropriate problems, outstanding integration issues, cognitive plausibility of the architectures, appropriate representational techniques, and the interaction with learning.

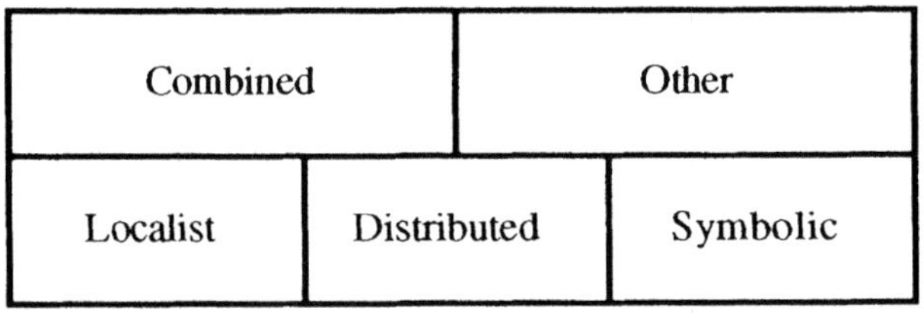

Figure 2.3 Architectural approaches to hybrid systems

Four basic architectures were identified, as shown in Figure 2.3. The localist approach implements symbolic structures in a network, e.g., each node representing a concept. This demonstrates the logical or rule-based capability of neural networks. The distributed approach is pure connectionism in which the network is functionally equivalent to symbolic processing. This is especially useful for natural language processing. The combined approach makes use of a set of separate modules with various degrees of coupling. Other architectures include the incorporation of neural networks into symbolic structures; the system is basically symbolic but has connectionist components.

Some major issues identified at the workshop include the interaction between the representation. Complex representations usually make learning more difficult. Guidelines are needed on what type to choose, how to build it, and how to know the constraints on learning. As listed in Figure 2.4, seven aspects of hybrid systems were identified as necessary for any model. Other criteria, shown in Figure 2.5, are important if the research is to produce fundamental advances in this field . Finally, Figure 2.6 illustrates the promising approaches as explained by Dreyfus. For more details on the ideas in these figures see Sun and Bookman, 1993.

frequency	things done often are done better
context prining	patterns processed earlier have significant influence
short-term vs. long-term	separate memory structures are needed
reduced representations	humans understand anyway
automaticity	things done often are done in abbreviated way
attention	as humans learn, representation goes from explicit to implicit
learning	system must learn from experience

Figure 2.4 Requirements for any hybrid model

Overall, two viewpoints seem to prevail. One is to integrate symbolic structures into connectionist architectures, using representations and techniques from each to tackle problems neither can solve alone; for example, modeling cognition that requires the ability to reason and the ability to perceive. The other viewpoint is to eliminate symbolic aspects altogether, and use only connectionist techniques.

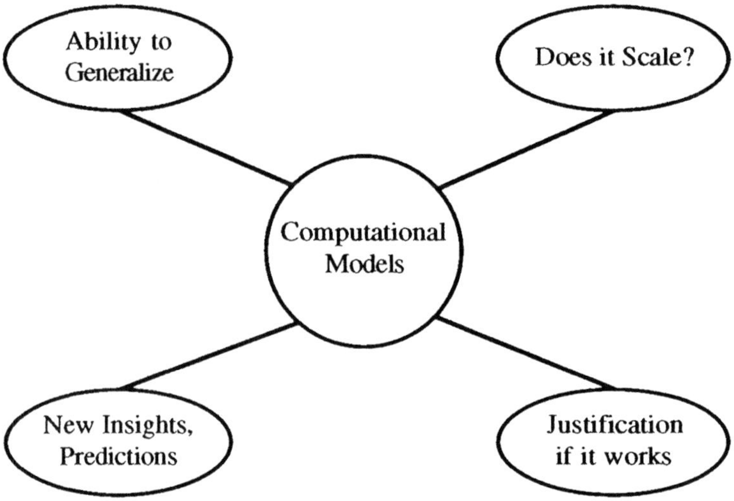

Figure 2.5 What to learn from hybrid system models (Hendler, 1992)

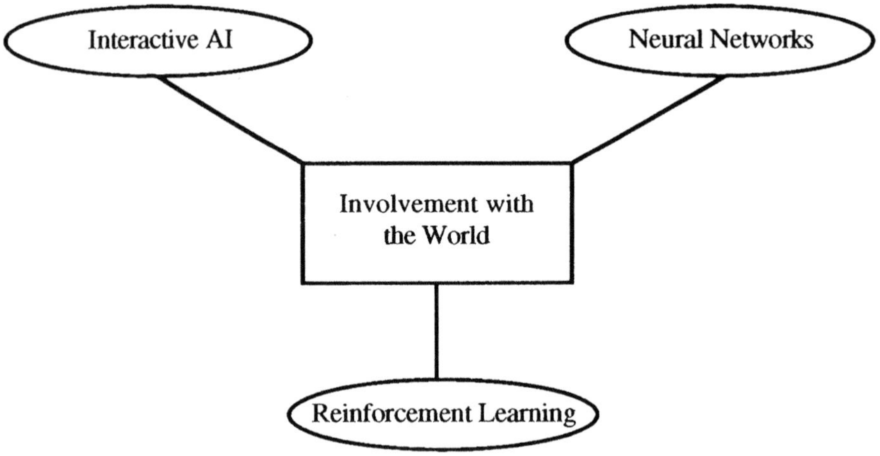

Figure 2.6 Promising approaches to intelligent systems

2.4 Development Tools

Several commercially available tools and environments are becoming available for important applications. However, the development systems are not yet as convenient and broadly useful as needed.

Research in this area looks at fundamental issues involved in linking symbolic and subsymbolic components and seeks ways to facilitate the creation of hybrid systems. A recent example of research in this area is the Conncert system described in more detail in Wilson and Hendler, 1993.

Conncert is a general purpose mechanism for linking hybrid system development with software development tools. This environment uses programming abstractions called supervisors that serve as intermediary agents between the symbolic and subsymbolic modules. A supervisor is bound to each subsymbolic network and functions as a monitor and controller of the network. Because of the high level interface, the details of the subsymbolic modules are hidden so that it can interact at the same level as symbolic components of the overall system.

In Wilson and Hendler, 1993, the following capabilities are identified for a system that links symbolic and subsymbolic components:

- where required, off-line training and tuning should be possible at various times during processing
- activation should be able to pass between the symbolic and subsymbolic components
- the capability should be available to translate subsymbolic output to symbolic representation
- both static and dynamic network creation should be supported.

Conncert, a framework and set of tools to provide the capabilities listed above, uses an intermediary mechanism (supervisor) to encapsulate the subsymbolic processing. This becomes a function is a program and provides the capability for abstraction. In Conncert, a supervisor is a software agent (written in C++) that provides an interface to a connectionist component and monitors its performance. Associates with each network, a supervisor provides an interface that allows high level information to be passed to symbolic components. The supervisor controls the environment (e.g., transitions between training and recall), inspects potential answers and passes them on if appropriate and legitimate, and provides an interface for requests to the network and its output back.

Figure 2.7 shows the general architecture for Conncert with the network encapsulated by the supervisor software. Two different information flows are provided: data into and out of the network and control signals that specify which corresponding function to perform. The main control loop of the supervisor is a large case statement with corresponding control signals.

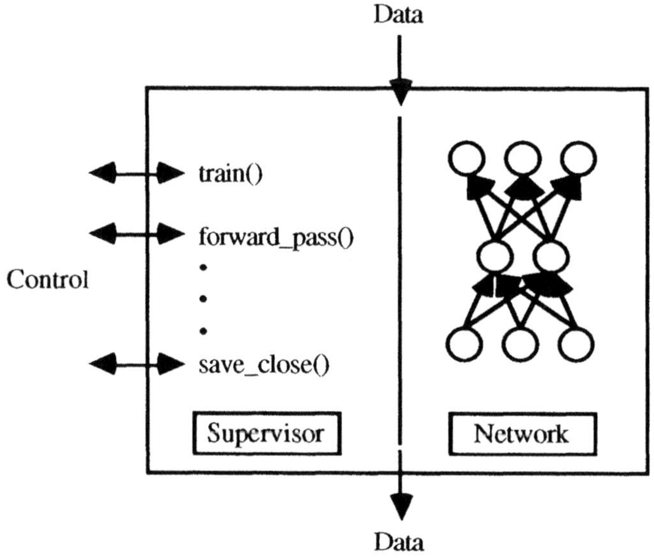

Figure 2.7 Overall architecture of the Conncert development system.

Supervisors are frameworks for monitoring the network, rather than being model based. The system creates function definitions as templates to which the user supplies contents specific to the desired network (e.g., Hopfield or backpropagation).

In addition to the modularity provided by encapsulation, the modules can execute in parallel. Interactions such as high level control requests are achieved via message passing, and objects can be interconnected to pass data between them. Furthermore, objects can be grouped and hierarchical components can be assembled to scale up the overall system.

Conncert is a framework for building potentially large systems from many interoperable symbolic and subsymbolic components. The key is the use of supervisors to raise the level of the connectionist model components. In particular applications, further programming is required to implement specific network models and to interpret their outputs for use by symbolic subsystems. Further research is needed to devise methods for monitoring networks and using the outputs. Also, more work is needed to apply this framework to highly integrated hybrid systems.

2.5 Status and Direction of Research and Development

Due to several promising applications and advances in neural network technology, considerable research and development activity is turning to ways of integrating appropriate intelligent system technologies. Hybrid systems have the potential of being more powerful than the systems built only with one of the component techniques.

Research on hybrid systems includes the study of obstacles to and solutions for the representation of symbolic and subsymbolic models so that they can function together. This requires mechanisms for communicating data and control signals and developing strategies for monitoring and controlling the symbolic and subsymbolic components. This includes issues in distributed AI for communication protocols among cooperating intelligent agents. Other work in hybrid system involves the creation of useful tools and development environments and the identification of the range of appropriate applications for hybrid systems.

Marvin Minsky (Minsky, 1991) has identified the need to build AI systems from a variety of components. AI researchers have tended to develop specialized, rigid models for narrow problems. Early attempts to make general problem solving models, while leading to fundamental progress, were not practical. The specialized systems produced many successful applications, but single models are too inflexible to meet the demands of real-world problems.

Although AI researchers have several goals, the field should ideally produce systems that learn from experience or from experiments and solve problems that we just pose to the machines. This often will require one or more different abilities we associate with human intelligence, including children's skills and commonsense reasoning, which so far have eluded AI.

"AI research must now move from its traditional focus on particular schemes. There is no one best way to represent knowledge or to solve problems, and the limitations of current machine intelligence largely stem from seeking unified theories or trying to repair the deficiencies of theoretically neat but conceptually impoverished ideological positions... The versatility that we need can be found only in larger-scale architectures that can exploit and manage the advantages of several types of representations at the same time. Then, each can be used to overcome the deficiencies of the others." (Minsky, 1991, p. 36)

Chapter 3

Models for Integrating Systems

Several techniques for integrating expert systems and neural networks have emerged over the past two years, ranging from the primarily independent to the highly interactive. While there are different approaches to categorizing these integration techniques, section 3.2 classifies them according to their software architecture.

3.1 IRIS Model

An overall framework for discussing hybrid systems can be found in Soucek (1991): Integration of Reasoning, Informing, and Serving (IRIS). As shown in Figure 3.1, IRIS is an architecture for combining software, hardware, and system levels involved in intelligent systems. This viewpoint is described as the key to more efficient technology, products, and services to meet business needs.

The systems approach of IRIS has as critical issues the engineering methods, tool sets, and interdisciplinary cooperation. Thus the development of hybrid systems integrates scientific disciplines including biology, cognitive psychology, linguistics, epistemology, and computer science. The importance of being able to process common knowledge and do commonsense reasoning is also acknowledged.

IRIS identifies the need for ten ingredients of integration:

- mixing of technologies (ES, NN, etc.)
- paradigms for integration
- standard software modules
- special languages
- software development tools and environments
- automated discovery such as interactive intelligent databases and interfaces
- standard control and automation modules
- case studies of working applications
- concurrency - tools for developing and monitoring
- signal-to-symbol transformations and pattern to category mappings - this includes facilities for sensor to ES, NN to postprocessing, and data to high level descriptions.

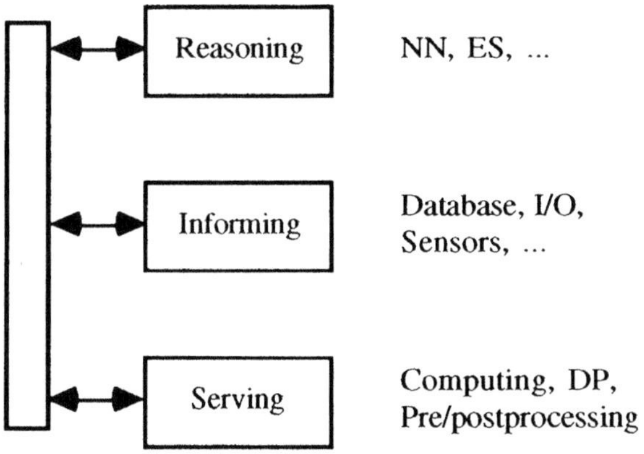

Figure 3.1 The IRIS architecture.

The IRIS framework also addresses the levels of integration, as shown in Figure 3.2. Current research and development has achieved some successes up to the cognition level, and some progress is starting to be made toward the highest level of conscience. This will require a new class of integration media including facilities for intelligent dialogues, intelligent message quanta, use of context, intelligent response and transfer functions, and fuzzy receiving and sending windows. This will be necessary if computers are to approach the human capabilities of thinking, reasoning, judging, and dealing with fuzziness and uncertainty.

3.2 Models for Integration

Five different integration strategies have been identified (see Figure 3.3): stand-alone models, transformations, loose coupling, tight coupling, and full integration (Medsker and Bailey, 1992). The following sections discuss each of these strategies, providing basic concepts and descriptions, an application example, variations and expected uses of the model, and benefits and limitations of the approach. Application examples are representative of, but do not describe actual integrated systems.

Connectivity	Telecommunication technology and infrastructure
↓	
Compatibility	Portability of tasks among computers
↓	
Coordination	Purpose of integration: better work, productivity, quality
↓	
Cognition	Integration of knowledge and intelligence from many sources; ability to reason, categorize objects into classes; recognize, associate, learn, generalize
↓	
Conception	Automatically and adaptively create new observables; e.g., artificial life
↓	
Conscience	Noble applications, good/evil; global thinking

Figure 3.2 Levels of integration

3.2.1 Stand-Alone Models

Stand-alone models of combined expert system and neural network applications consist of independent software components. These components do not interact in any way. While stand-alone models are a degenerate case for integration purposes, they are an alternative worth discussing.

Several purposes exist for developing stand-alone expert systems and neural networks. First, they provides direct means of comparing the problem-solving capabilities of the two techniques for a specific application. Second, used in parallel, the techniques provide redundancy in processing. Third, developing one technique after finishing a model of the other facilitates validating the prior development process. Finally, running two models in parallel permits a loose approximation of integration.

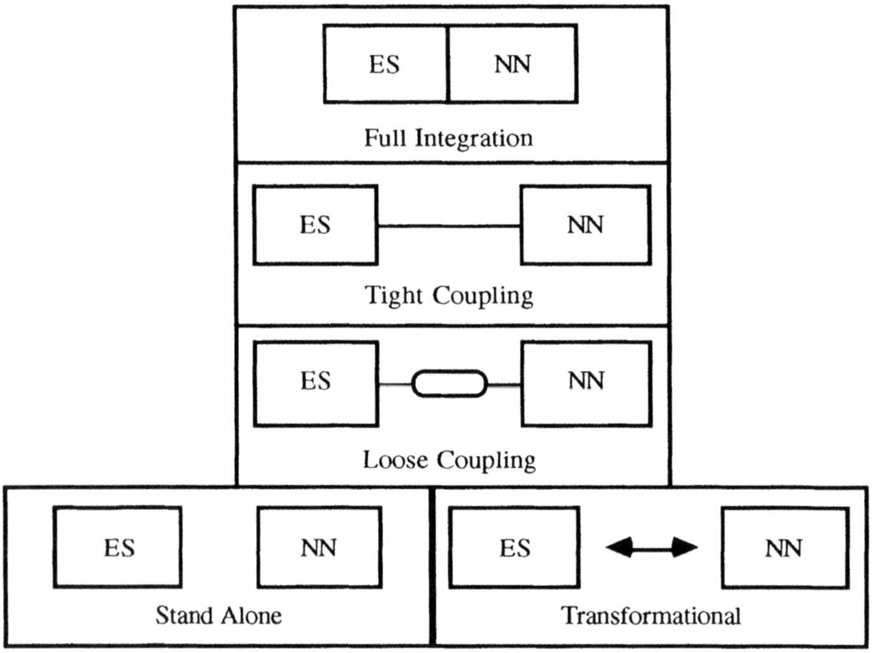

Figure 3.3 Models for integrating expert systems and neural networks

An example of a stand-alone expert system/neural network model involves the diagnostic classification of symptoms in computer repair. Two distinct components, an expert system and a neural network, are developed to solve the same classification problem. When a computer malfunctions, symptoms are presented to both the expert system and the neural network and both return a solution. The independent solutions are compared, and if there is disagreement, the user selects which solution to implement. The expert system can also be queried for justification for its result.

Stand-alone models serve three primary purposes. First, as parallel systems they offer two sets of capabilities: the generalization and adaptability of neural networks, and the stepwise deduction and explanation facilities of expert systems. Second, stand-alone models provide verification of previous applications. Oftentimes, developers will build a neural network to solve the same problem as an existing expert system. This permits the developers both to compare the capabilities of expert systems to those of neural networks, and to ensure that the initial system performs properly.

Finally, stand-alone models are used to quickly develop an initial prototype, while a more time-consuming application is developed. For example, a neural network might be trained to solve a problem temporarily while a more complete expert system solution is developed. A rapid prototype of either a neural network or an expert system provides two benefits. First, it provides a quick problem-solving tool that can satisfy short-term needs while the full-scale system is developed. Second, the experience of conducting the initial development process often provides important guidance for the full-scale system by highlighting requirements and pointing out pitfalls.

Stand-alone models have two principle benefits when compared to other forms of expert system/neural network models. First, because they do not attempt to interface with each other, the model is straightforward to develop. Second, there are no impediments to the use of commercially-available software packages.

On the other hand, stand-alone models have several limitations. Other than conceptual issues, there is no effective means of leveraging the development efforts of one technique when developing the other. Neither the neural network nor the expert system can support the weaknesses of the other technique. The systems are completely independent and their benefits are simply those derived from their separate technology. Finally, developing separate systems effectively doubles the maintenance requirements for the model. Both must be updated simultaneously to avoid confusion, and updates to one cannot help the other.

3.2.2 Transformational Models

Transformational models are similar to stand-alone models in that the end result of development is an independent model that does not interact with another. What distinguishes the two types of models is that transformational systems begin as one type of system (e.g., a neural network), and end up as the other (e.g., an expert system).

As indicated in Figure 3.4, transformational models have two forms: expert systems that are transformed into neural networks, and neural nets that metamorph into expert systems. Determining which technique is used for development and which is used for delivery is based on the desirable features that the technique offers.

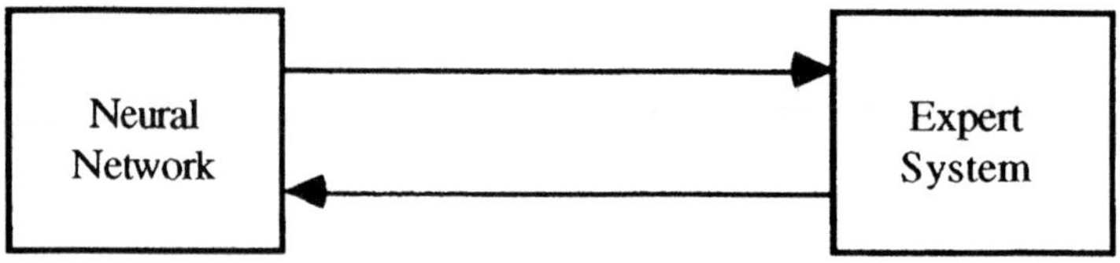

Figure 3.4 Transformational model for hybrid neural and expert systems.

An application example of a transformational model is a marketing decision aid. Initially, a neural network is developed to identify trends and relationships within sales data. Then the neural network is used as the basis for an expert system that assists marketing researchers in allocating advertising resources. In this example, the neural network is used to quickly adapt to a complex, data-intensive problem, to provide generalization, and to filter errors in the data. An expert system was targeted as the delivery system because of the desire to document and verify the knowledge used to make decisions, and because the users required justification capabilities.

Neural networks that are transformed into expert systems are often used for much the same purpose as described in the example above. Data analysis and preliminary knowledge engineering are principle applications for this type of transformational model. The neural networks are transformed into expert systems for reasons such as knowledge documentation and verification, the desire for stepwise reasoning, and for explanation facilities.

While less common, the expert system to neural network transformational model is also useful. The expert systems are usually converted for one of two reasons. Either the expert system was incapable of adequately solving the problem, or the speed, adaptability, and robustness of neural networks was required. Knowledge from the expert system is used to set the initial conditions and training set for the neural network, and the neural network evolves from there.

Transformational models offer several benefits to developers. They are often quick to develop and ultimately require maintenance on only one system. Development occurs in the most appropriate environment. Similarly, the delivery technique offers operational benefits suitable to its environment.

Limitations to transformational models are significant. First, there is no fully automated means of transforming an expert system to a neural network or vice versa. In fact, there is no known method for accurately and completely performing the transformation. However, the fact that transformational models are relatively common demonstrates that adequate transformations are possible with reasonable resources.

Another limitation is that significant modifications to the system may require a new development effort, which leads to another transformation. In addition to maintenance issues, the finished transformational system is limited operationally to the capabilities of the target technique. Thus, the benefits of integrated systems are not truly enjoyed.

3.2.3 Loosely-Coupled Models

Loosely-coupled models are the first true form of integrating expert systems and neural networks. The application is decomposed into separate neural network and expert system components that communicate via data files (Figure 3.5). Among the variations of loosely-coupled models are preprocessors, post-processors, co- processors, and user interfaces. For the purpose of this discussion, we will consider the neural network component of the model to be the pre- or post- processor.

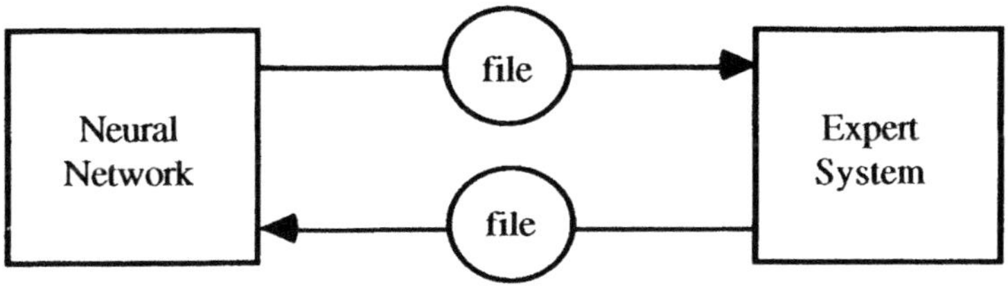

Figure 3.5 Loose-coupling model of hybrid neural and expert systems.

For an application example, consider a model forecasting the utilization of a work force. Data is fed into a neural network that predicts the workload for a given time period. The forecast is placed into a data file, and passed to an expert system that uses the workload to determine the utilization of the workforce.

In preprocessing loosely-coupled models, the neural network serves as a front-end that conditions data prior to passing it on to the expert system. Expected uses for this type of model include using the neural network to perform data fusion, to remove errors, to identify objects, and to recognize patterns. The expert system component can then use this information to solve problems in classification, identification, scene analysis, and problem solving.

Post-processing models are the converse of preprocessing models. In post-processing, the expert system produces an output that is passed via a data file to the neural network. In this type of architecture, the expert system can perform data preparation and manipulation, classify inputs, and make decisions. The neural network component then performs functions such as forecasting, data analysis, monitoring, and error trapping.

The co-processing model involves data passing in both directions, which allows interactive and cooperative behavior between the neural network and expert system. While very few co-processing applications are available, they have the potential for solving difficult problems. Possible applications include incremental data refinement, iterative problem solving, and dual decision making.

User interfaces are turning to neural networks as a pattern recognition technology capable of increasing the flexibility of user interactions with expert systems. Initial research often takes the form of loosely-coupled models, which allow projects to focus on pattern recognition rather than integration issues. Speech processing and handwritten character recognition are perhaps the most common forms of user interfaces, but image processing and user modeling are also under research.

Compared to the more integrated expert system and neural network applications, loosely-coupled models are easy to develop. They are amenable to the use of commercially available expert system and neural network software, which reduces the programming burden on the developers. Both the system design and implementation processes are simplified with loosely-coupled models. Finally, maintenance time is reduced, because of the simplicity of the data file interface mechanism.

Four limitations are associated with loosely-coupled models. First, because of the interface, operating time is longer for loosely-coupled applications. Second, there is often a great deal of redundancy in the development of the separate neural network and expert system components. Both must be capable of solving subproblems in order to perform their unique computations, but because they lack direct access to each other's internal processing they must develop independent capabilities. This also leads to overlap in the data input requirements and internal processing. Finally, there is a high communications cost for loose coupling.

3.2.4 Tightly-Coupled Models

The categories of loose and tight coupling have significant overlap. Both utilize independent expert system and neural network components. However, tight coupling systems pass information via memory resident data structures rather than external data files. This improves the interactive capabilities of tightly-coupled models in addition to enhancing their performance.

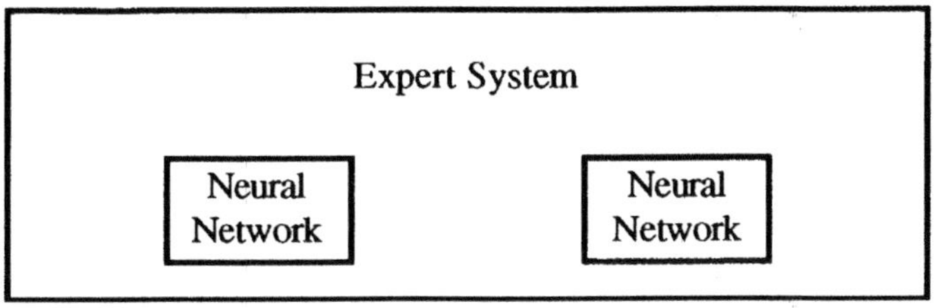

Figure 3.6 Tight-coupling model of hybrid neural and expert systems

Tightly-coupled models can function under the same variations as loosely-coupled models, except that the tightly-coupled versions of pre-, post-, and coprocessors are typically faster. Variations unique to tight coupling include blackboards, cooperative, and embedded systems.

Another forecasting application provides an example of tight coupling. In this instance, stock option data is presented to a neural network. The network uses financial and stock option data to predict the options strike price over a three-day period. This information is then passed to the expert system which determines the appropriate unwind (action) strategy for the option.

One of the most interesting expected uses of tightly-coupled models is in the area of blackboard architectures. Blackboards are shared data structures that facilitate interactive problem solving via independent agents. Typically the agents are knowledge-based systems. It is both technically feasible and operationally important to consider the potential of adding neural networks as agents to the blackboard paradigm. Applications for integrated blackboard systems include complex pattern recognition, fault isolation and repair, and advanced decision support.

Cooperating systems are one of the most common variations of tightly-coupled expert system/neural network models. Cooperating systems are similar to co-processing loosely-coupled models but tend to be highly interactive due to the ease of data-passing. Applications of cooperating systems occur in monitoring and control, decision making, and several problem solving domains.

Embedded systems are a third variation of tightly-coupled models that use modules from one technique to help control the functioning of the other technique. For example, neural networks can be embedded inside expert systems to control the inferencing process. Embedded neural network components are used to focus the inferencing, guide searches, and perform pattern matching. Expert system components can be used to interpret the results of neural network, to provide internetwork connectivity, and to provide explanation facilities. Applications of embedded systems exist in the areas of robotics, education, and classification.

Tight coupling has the benefits of reduced communications overhead and improved runtime performance, when compared to loose coupling. By maintaining the modularity of the expert system and neural network components, several commercial packages are suitable for developing tightly-coupled models. Overall, tight coupling offers design flexibility and robust integration.

Tightly-coupled systems have three principle limitations. First, the development and maintenance complexity increases due to the internal data interface. Second, tight coupling suffers from redundant data gathering and processing, just like loose coupling. Once again, this is due to the independence of the expert system and neural network components. Finally, the verification and validation process is more difficult, particularly for embedded applications.

3.2.5 Fully-Integrated Models

Fully-integrated expert system/neural network models share data structures and knowledge representations. Communication between the different components is accomplished via the dual nature (symbolic and neural) of the structures. Reasoning is accomplished either cooperatively or through a component designated as the controller. Several variations of fully-integrated systems exist, including connectionist systems, the utilization of I/O nodes, subsymbolic to symbolic connectivity, and integrated control mechanisms.

For an application example, consider a fully-integrated expert system/neural network that identifies objects. The neural network receives feature data from sensors and environmental data from the expert system. The neural network produces a preliminary assessment of the object, which the expert system uses to further refine the working hypothesis. Once sufficient evidence has been gathered to support the hypothesis, a solution is presented to the user. Communications in this example are accomplished through the sharing of nodes and symbols. Input and output nodes from the neural network are also used as symbols by the expert system. Information is passed back and forth by changing the values and activations on these dual structures.

The most common variation of fully integrated models is the connectionist system, and more specifically connectionist expert systems. Connectionist systems in general often rely on local knowledge representations, as opposed to the distributed representation of most neural networks, and reason through spreading activation. Connectionist expert systems represent relationships between pieces of knowledge with weighted links between symbolic nodes. Applications of connectionist expert systems exist in medical diagnosis, information retrieval and analysis, and pattern classification.

Utilizing the input and output nodes of a neural network as facts within an expert system is a second variation of fully integrated models. This allows the expert system and neural network to interact quickly and easily. Information is available to either component instantly, and it is common to approach problem-solving tasks incrementally. Applications of utilizing I/O nodes exist in diagnosis, pattern recognition, and classification.

A third variation of full integration involves linking subsymbolic to symbolic computing. This is accomplished by connecting nodes and patterns of activation within the hidden layer of a neural network to symbols within an expert system. This effectively links the distributed representation developed by the neural network training process to the local representation of an expert system. The process of connecting internal nodes to symbols is often based on both the analysis of the weighted links within the neural network, as well as the application of statistical clustering techniques. There are several objectives to this type of integration. One is to study the nature of distributed representations and how the brain might store information. Another is to access more detailed information from the neural network than is presented at the output nodes. Applications include image processing, feature extraction, and decision making.

The final variation of fully integrated systems is integrated control, which is related to the embedded tightly-coupled systems. Integrated control differs in that the expert systems and neural networks are no longer maintained as independent modules. Instead, processes and data are shared as much as possible to minimize redundancy in development and in operation. Applications of integrated control include focusing inference, selecting among hypotheses, controlling agendas, and providing search heuristics.

The benefits of full integration include robustness, improved performance, and increased problem solving capabilities. Robustness and performance improvements stem from the dual nature of the knowledge representations and data structures. In addition, there is little or no redundancy in the development process, because the systems can leverage off of each other. Finally, it has been demonstrated that fully integrated models can provide a full range of capabilities -- such as adaptation, generalization, noise tolerance, justification, and logical deduction -- not found in non-integrated models.

Full integration has limitations caused by the increased complexity of the inter-module interactions. First, there is the complexity of specifying, designing, and building fully-integrated models. Second, there is a distinct lack of tools on the market that facilitate full integration. Finally, there are important questions in verifying, validating, and maintaining fully-integrated systems.

3.3 Summary

Practical applications of hybrid neural network and expert systems have only recently started to be used. As more experience is gained better models, methods, and procedures will be discovered. As described in Part II of this book, case studies of successful systems are raising the awareness of the potential of hybrid systems. With better guidelines, a broader range of applications can be developed in a more reliable, efficient manner. The IRIS viewpoint is a reminder that hybrid neural network and expert systems need also to be developed in the context of even higher levels of integration.

Part II Case Studies of Hybrid Neural Network and Expert Systems

Chapter 4

LAM^{tm} Hybrid System for Window Glazing Design

The LAM^{tm} hybrid system is currently in use by architects, glazing specifiers, and laminators. LAM^{tm} was developed by Ray V. Foss, a senior research engineer with E. I. du Pont de Nemours & Company, Du Pont Polymers and Automotive - "Butacite", Parkersburg, W.V. Dieter H. Droste also collaborated in this project. Mr. Foss is the primary author of this chapter.

An integrated neural network and expert system was developed to facilitate the design of window glass for structural strength, hydrostatic loads, sound attenuation, and solar control. The integrated software consists of a text interface, rule-base expert systems, and two neural networks. The text of a glazing design guide is linked by related topics and concepts. The user enters design parameters for the window choice via an interactive consultation into a rule-base expert system that critiques the design.

Technical aspects for the glass structural strength are based primarily on linear methods published by the American Architectural Manufacturers Association (AAMA). Statistical correlations for the new non-linear failure prediction for glass strength are used to design automatically for minimum glass thickness required to withstand a specified load. Neural networks estimate values for sound attenuation and solar transmission characteristics from laboratory test data on selected glass constructions.

Combining rules with neural nets illustrates an effective marriage of formal logic (rules) with plausible reasoning (neural nets). This hybrid system illustrates the potential for not only integrating a broad scope of diverse but related knowledge, but also for facilitating the important, indeed crucial, nature of the communication processes inevitably involved.

Development of LAM^{tm} glazing design software was originally intended to assist glazing design in a multi-faceted, comprehensive and integrated sense. But, in retrospect, the processes of communication involved in all aspects of the system - from knowledge acquisition and development to system implementation and client interactions - are by far the most valuable and instructive result of the system in practice.

4.1 Laminated Glass and Window Glazing Design

The glass design software (LAM, 1989) is a hybrid expert system which helps users design a variety of window glass constructions for buildings. This expert system was inspired by the standards manual published by the laminators trade association (LSGA, 1985). But the software automates strength design procedures using an approach consistent with model codes. The system scope is not restricted to glass laminates but covers common monolithic glass as well. The software determines the structural strength and estimates the sound control and solar properties for three types of glazing: (1) monolithic glass, (2) glass laminates and (3) insulated glass (IG) units.

This chapter describes how models of knowledge representation were combined into an integrated system. It presents some examples of strength calculations and of laminated glass property predictions to demonstrate the versatility and usefulness of an integrated expert system to handle standard glass design procedures. This expert system for glass design is one of the first design software tools available to combine rule-base expert systems with neural networks (Foss and Droste, 1990).

4.1.1 Background on Laminated Glass and Its Application

Laminated safety glass consists of two or more layers of glass, factory bonded together with a tough, flexible polyvinyl butyral (PVB) or other plastic interlayer. Plastic interlayer for automobile windshield glass lamination has been used since the 1930's. DuPont is one of two major global producers of PVB interlayer for laminated glass. While PVB is a well established commercial product, the manufacturers neither produce nor supply laminated safety glass itself.

Glazing is the only construction material which both separates and joins space to the senses. Increased security needs and consumer concerns about safety and energy efficiency are expanding the use of flat laminated glazing in the architectural market. In 1992, about 25 percent of the PVB interlayer produced worldwide was used in architectural applications. Architects around the world have increasingly turned to laminated glass for its safety and energy-saving features as well as its transparent beauty.

While the primary concern with laminated glass in automotive windshields is its impact resistance when broken, architectural laminates offer many advantages over monolithic glass for building glazing. Important safety properties include penetration and shatter resistance at all temperatures; no deterioration or discoloration from sunlight or moisture; extreme flexibility under impact; ease of cutting for replacement work; and perfect visibility. For architectural applications, other properties include: improved sound control; solar energy control; reduction of glare and fading; and resistance to windblown debris and wind or snow loads. Using multilayering techniques, architectural laminated glass can also be made to protect against theft, bullets, explosions, and physical attacks.

4.1.2 Attributes and Benefits of PVB Interlayer in Laminates

Laminated safety glazing offers PVB performance features important to residential and commercial architectural design, including:

Solar Control -- Laminated glass with tinted PVB interlayer saves energy by controlling heat buildup in summer, especially when used with absorbing or reflective glass. In winter, heat loss can be reduced, especially when used with insulated and low-E glass. Tinted interlayer also reduces glare and fading and screens out ultraviolet rays; greater solar control is obtained by specifying darker tint. However, in applications requiring both solar control and highly transparent glass for visibility (such as residential), a prototype "Butacite" PVB laminate used spectrally selective glass tint in combination with hard-coated pyrolytic low-E glass for both superb clarity and good shading and such may expand design choices for cool daylighting.

Reduction of Injury Risks from Broken Shards -- Laminated safety glass is particularly beneficial in overhead uses. If the glass is broken, fragments are likely to adhere to the PVB interlayer, reducing the risk of laceration and glass fallout. Using laminated glass for overhead glazing protects people from broken glass.

Improved Sound Control -- The PVB layer in laminated safety glass produces a damping effect to reduce noise transmission, providing an effective noise barrier across a wide range of frequencies. Airports, hotels, recording studios, and buildings near airports, highways and rail lines frequently rely on this feature of safety glass.

Security Against Theft, Bullets, Explosions, and Physical Assault -- By resisting penetration from hand-held and hand-thrown objects, laminated safety glass helps make homes, schools and offices more burglar resistant. Bullet resistant laminates also resist penetration from medium to superpower small arms and high-power rifles.

Resistance to Windblown Debris and Wind or Snow Loads -- In hurricane-prone areas, laminated glass has shown its ability to resist glazing failure from windborne missiles. This is particularly important for public buildings that must provide continued operations. When broken, laminates retain most glass shards and can maintain the integrity of the window opening.

Design versatility -- Laminated glass manufacture offers a wide variety of design options, including flat or bent glass; heat or chemically strengthened glass; and clear, tinted or reflective glass. PVB colors are light bronze,; grey; blue green; azure blue; light medium and dark brown; and clear. One designer has developed a process for encapsulating fabric between laminated glass plies.

Advanced Glazing Products for Intrusion or Architectural Graphics -- Three recent introductions include: "SentryGlas", a composite of "Butacite" film and strong, abrasion resistant polyester film for organic coated glass which controls broken glass particles and is particularly useful in areas where severe weather such as hurricanes are common; "ChromaFusion", a new technology to capture any texture, color or graphic image in laminated glass; "ContraVision", a laminated graphic glass with a color image which is clearly visible on one (exterior) side of the glass but optically disappears on the other (interior) side.

4.1.3 Structuring and Prioritizing the Knowledge Base

The benefits are many. But both design methods and machine representation thereof will lag behind the imagination of the designers. LAM^tm^, while integrated, is necessarily limited in scope. Laminates can be selected to avoid breakage from wind pressure. When architectural laminates break, the PVB interlayer is important for safety, but it also serves functionally for sound reduction and, if colored, for solar absorptance. This much was represented. The system illustrates integrated design for a hierarchy of multiple functionalities.

Integration has the advantage of producing a functional design across multiple criteria, while requiring less designer time. Modularization addressed maintainability issues.

Integration -- The scope of LAM^{tm} is an integrated approach to window glazing design across three separate but related glass functionalities: 1) pressure strength, 2) solar control, and 3) sound reduction. LAM^{tm} is designed to interact with the user to select a glazing meeting target values for these functions; for strength, the target is whether or not the glass is acceptable by established standards for a given wind load, whereas the other targets are user defined. LAM^{tm} assumes the window is a system isolated from the context of larger systems, like the building and its environment. The conceptual user interface is one of interaction between the software and the user, a question and answer paradigm, advising on choices preferred by the user. LAM^{tm} does not produce new options or comprehensive lists of possible user choices.

Hierarchy of Functionalities -- A primary technical or knowledge engineering concept used in developing LAM^{tm} is a perceived hierarchy of importance among the knowledge topics represented, specifically the glazing functionalities. Glass strength was more important than sound and/or solar control. But as the system was discussed with clients and knowledgeable users, it became apparent that not only was the technology around glass strength in a rapid state of development and change, but the solar control functionality is probably much more important than its representation in LAM^{tm}, which describes only solar properties of laminated glass with tinted PVB plastic interlayers, a very restrictive subset. Even though LAM^{tm} is broad in scope and depth on glazing wind load strength and thickness selection, emphasis was on laminated glass properties and applications.

Modularization -- A key concept used in programming the knowledge representation was linking modules. The glazing design functionalities were separated into modules for strength, sound and solar. Integration is achieved not only by the structure of the system but by passing a file of glass specifications and "facts" from module to module. Modularization has major advantages to practical system maintenance, but makes the system performance rather algorithmic, like subroutines.

4.1.4 The Task of Manually Designing Window Glass

First and foremost, glazing for architectural applications must have adequate structural strength to safely meet the building and code requirements (see

Figure 4.1). A serviceable and acceptable design should not break and should not deflect excessively under wind loads or, if installed non-vertically, under snow and live loads. Calculation of glass structural performance is basically straight forward, but the various code and recurrent design requirements make it a wearisome task. The designer has to consider which of the different design equations are applicable, and what glass design factors are required for particular installations.

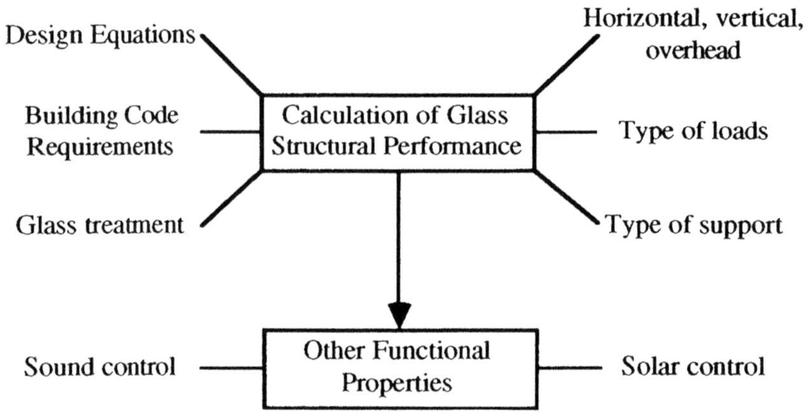

Figure 4.1 Tasks in the design of window glass

Design methods for most vertical applications require a breakage probability level of 8 lites per 1000 (0.008) or less, but glazing installed overhead must meet a stricter probability level of 0.001 or less. The type of support for the glass unit, i.e. whether it is two or four-sided, will affect the design calculations. Choices for glass treatment, which include annealed "Kind AN", heat strengthened "Kind HS" or fully tempered "Kind FT" (AAMA, 1988), will determine its structural strength. Laminates or the kind of glass treatment allowed in some particular applications is governed by building codes.

Once the glass construction and its thickness have been determined for adequate structural strength, the problem becomes one of estimating other functional properties, such as sound and solar control. The estimation of these properties for the selected glass unit is somewhat more difficult. This information is generally only available in the form of tabulated test results, and quite often tables do not list the particular glass construction the user is interested in. The correlation of solar and sound properties are fairly complex and a simple interpolation of available test data is not practical. Often the designer is left with making an educated guess.

4.1.5 Technical Details and Definitions of Terms in Window Glazing Design

Since work on this expert system began, various aspects of the glass design basis have been changing. Design standards for strength are changing from the older, linear or "straight line log log" methods to the newer, non-linear methods which consider both aspect ratio and size effects of the glass. The new failure prediction model is based on a finite element mathematical model that predicts load lines for glass sheets at a given breakage probability. Load duration effects reduce the strength of glass under sustained stress; this may be accounted for by increasing factored load using an "equivalent uniform load" concept described in some model codes or by applying a factored glass strength reduction.

It is well known that glass is a brittle, "statistically variable" engineering material; its strength varies markedly between nominally identical panes. Many factors affect average breakage strength of glass under uniform pressure: thickness; size or area; heat treatment or chemical strengthening; shape, namely aspect ratio and width-by-thickness or "flexibility" ratio; time at load or dynamics; surface flaws and aging; edge flaws and cutting; edge restraints, and; assemblies such as IG units or laminates. Primary affects on average breaking pressure are thickness and size. The larger the glass plate or the longer the load acts, the lower the average breaking strength, for a given thickness and heat treatment (Calderone and Melbourne, 1993).

An acceptable architectural glazing must have a design strength greater than the applied load. "ACCEPTABLE GLASS DESIGN" means the primary criteria for structural strength is met. That is, factored glass strength must be greater than factored wind load. The basis for structural adequacy in design methods is founded ultimately on actual tests to breakage in which "failure strength" of newly produced glass, supported at its edges, was exposed to uniform and increasing loads until breakage. The definition of failure has been breakage of any glass due to uniform pressure. This assumes a paradigm of minimizing risk of glass shattering. Such glass fracture or cracking in multi-layer assemblies may not necessarily result in violation of the building envelope, a concept largely discounted in current laminate selection practices.

So "design strength" is simply the breakage "failure strength" divided by a "glass design factor", usually 2.50 for a normal breakage probability, giving 8 failed lites per 1000. Once the design for strength is "ACCEPTABLE", less critical design criteria should be serviceable, for deflection, sound and solar transmission. So the user must interact with the system within the function of interest, which has advantages and disadvantages. Users can iterate sequentially on glass design choices to reach the defined targets within each function, strength or sound or solar. Since the modules are relatively independent, the user must process each function separately.

A much debated issue in the industry is the strength of a glass laminate relative to monolithic glass at the same total thickness. Concepts used for laminate strength predictions include the "layered model" in which the laminate acts like two glass plies separated by a hypothetical fluid (Sowers, 1981) or the "monolithic model" where laminated strength is equal to that of equivalent total glass thickness (CGSB, 1989). Stress theory mathematical modeling has shown large, thin laminates can approach, if not exceed, the strength of monolithic (Vallabhan *et al.*, 1991).

Strength estimates of glass laminates range from the monolithic model to the layered model. Recent design methods establish limits as to when these models apply, such as for elevated temperatures which affect the physical properties of the PVB interlayer. These limits have been incorporated in a new non-linear method (CGSB, 1989). The older, linear design methods for laminate strength require the strength of the equivalent total monolithic glass thickness to be multiplied by a design factor (usually less than 1.0). Since a value of 0.75 for the laminate design factor has been sanctioned (AAMA, 1988) for use with the linear design methods and is standard practice, this was advised in this system, but has been made user definable (see section 4.4.1 and Table 4.5).

4.1.6 Significance of the Hybrid System to the Knowledge Engineering Process

The task of designing architectural glass units fits well into the realm of an hybrid expert system, because of its advantages to easily retrieve text information (such as design references), to quickly make logical decisions and to accurately perform arithmetic computations (for structural design calculations), and to intelligently estimate complex correlations (such as those for sound and solar properties).

But an important consequence of developing LAM^{tm} is the value of the communication processes involved in the system's knowledge acquisition and representation.

The influence of LAM^{tm} on the communication processes cannot be understated, but may certainly be non-obvious. At a level of knowledge beyond that actually represented, there is the process of creating new knowledge, enhanced by the relationships between the system developers and industry glass experts. There are multiple examples, the most significant involving participation by the system developers on industry wide glazing standards or methods development organizations. This collaboration essentially expanded the glazing knowledge base itself. So in even a larger sense, the LAM^{tm} hybrid system is both a product of a kind of human-machine interaction, and has been a catalyst for enhancing glazing knowledge development. The challenge is cost effective system maintenance.

This application is significant at several levels. Internally within DuPont, it provides a technical marketing communication link from the "Butacite" business to both the direct customer base as well as the level beyond, the end users or customer's customers. Externally from the users viewpoint, the system provides interactive ease of access to technical performance properties of glazing systems, saving time in glass selection. But the perceptual or conceptual significance lies in the valuable communication processes created by the development of a system broad enough in scope to facilitate interactions among the people involved.

Hybrid systems can be a means to the larger end of nurturing and managing the communication processes involved in knowledge acquisition and representation.

4.2 Architecture, Design, and Development of LAM^{tm}

The contents of this expert system were assembled from multiple sources and from published industry standards (Oxman and Dockery, 1990). The novel feature of the hybrid system was to combine three different software shells into one, easy to use program: text, rule-base systems and neural networks. The subsystems were designed to handle three different tasks, to retrieve and display text information, to perform a strength analysis of the selected glass constructions, and to estimate the sound and solar properties of the chosen design (see Figure 4.2).

The text interface allows the presentation of conceptual material to the user. The text consists of the complete contents of a glass design manual, including its glossary, trade names and member directories. The text modules allow users to view the manual on screen and link interactively to concepts and pertinent cross references.

All logic decisions and numerical calculations were handled by the rule-base system. The calculations involved primarily the structural strength and

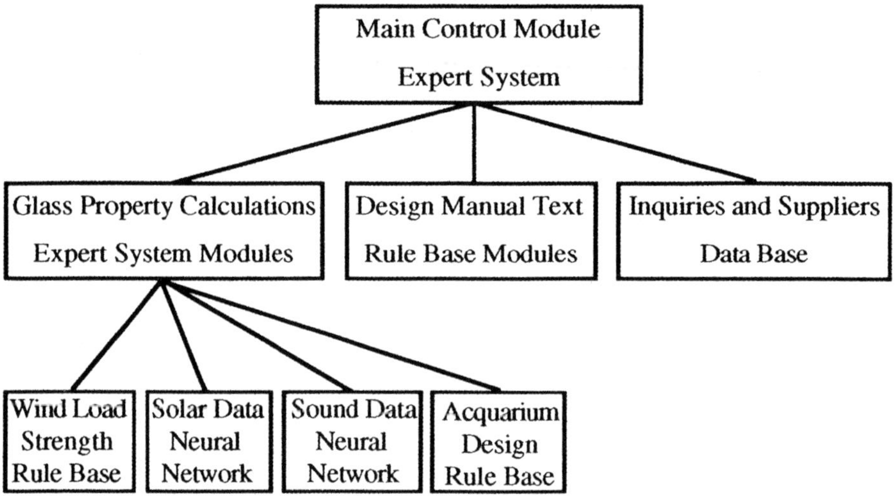

Figure 4.2 Architecture of the LAM hybrid system

the deflection of the glass construction as well as its breakage probability. The rule-base system also defines the type of glass unit and its construction by presenting menu choices to the user to facilitate the user's selections. Another function of the rule-base system was to pre-process the inputs to the neural nets, thereby eliminating a direct interface between the user and the neural net. This not only reduced input errors, but also permitted the neural net to be hidden from the user.

Transparent to the user, the glass design expert system employs a secondary expert system, neural network, which had been trained with published test data for solar and sound properties. When the properties for a particular glass construction are requested, the neural net recalls them from the learned data or, if they did not exist as test data, the net makes a "best estimate" of those properties, based on inferences from the other available glass constructions in the test set. The neural net acts as an efficient substitute for a large number of explicit rules or data set interpolations, and the net can generate estimates for cases not explicitly defined in the training set.

The glass design hybrid system was designed as 16 independent modules linked with one common 'fact file' to pass all computational results and control parameters from one module to the next. The interface between the rule-base expert system and the neural network allows if/then rules to pass a structured query to the neural net and permits further inferences by the rule-base system on the information returning from the net.

The combination of neural nets with rule-base expert systems has several advantages. The rules formulate a validated and structured input to the network that is presented in the same pattern in which the net was trained. This permits reliable use of nets with multiple or mutually exclusive inputs. This novel combination of expert systems and neural nets offers a new approach to the way glass constructions are designed, and new capabilities for representing many kinds of glass laminate performance.

All three tasks described above are accomplished with commercially available software for DOS compatible personal computers. Customized screens, generated by separate PASCAL programs, were employed to let the user enter, modify or correct all inputs for a consultation on one interactive screen (Figure 2). The system is initialized with the design case previously computed and uses it as a template for the current consultation. The complexity of integration made verification a challenge, but once the expert system was organized, incremental knowledge, such as prediction of other properties for glass constructions, can be easily added.

4.2.1 Design of the Expert System Component

Thirteen of the modules were written in a production rule language and contain a total of 578 if/then rules:

(1) the main control module for the system that performs the rule-base inferences, classifies glass types, makes logic decisions, and checks for errors or improbable glass designs;
(2) four modules that perform numerical calculations of glass properties; and
(3) those modules that present menu choices and execute external programs, such as screen displays, neural networks or data base interfaces. The advantage of the rule-base expert systems is that all their logic conclusions are exactly defined. They do not make inferences unless there is an explicitly defined rule. Rule-base systems handle arithmetic and logical computations with ease. The use of such a system within the glass design software also simplified the many required inputs by presenting the user with various menu choices. Once the user's choice is indicated by a single selection, the expert system transfers the pertinent

numerical or logical values (in this software, one selection can enter as many as five parameters) into the "fact file" (see Figure 4.7). This method of inputting parameters speeds up data entry and reduces typing errors.

The design rules for the structural strength of glass were constructed from published straight line, log-log relationships (AAMA, 1988). Glass strength equations are defined by "K4" values for four-sided support, with area limits depending on the particular glass construction. These "K4" values, the product of design pressure times design area for each glass thickness, express the relationship between allowable glass area and maximum design pressure. Maximum area is inversely proportional to allowable design pressure. Larger plates are weaker and can resist less pressure: K4 = constant = pressure x area.

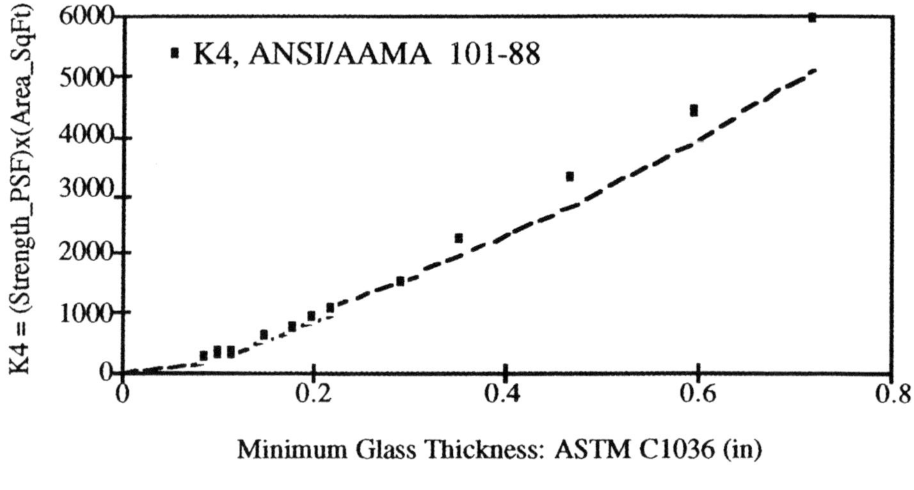

Figure 4.3 Glass strength represented as pressure-area constants

Acceptable glass selection, then, is based on the array of "K4" constants for each glass thickness, varying exponentially by about the 1.7 power of thickness (AS 1288). Each K4 value corresponds to both a particular glass thickness and a series of rules that not only describe strength, but also apply practical area limits for that thickness (see Figure 4.3). Predicted glass strength based on the Australian Standard appear to be about 12% more conservative than the AAMA Standards. The glass thickness for a given external load and its estimated deflection were computed from equations published by Canadian General Standards Board (CGSB, 1989).

4.2.2 Neural Network Component

Neural nets were useful for design interpolations. Neural nets appear practical for effective representation of design data, when a sizeable training set exists that correlates the functional relationships.

Neural networks are based on pattern recognition and have an advantage over rule-base expert systems when either the rules for describing a particular system problem become too complex or when the problem cannot be described with rules at all (Lippmann, 1987). The knowledge contained in neural nets is not coded by the systems analyst or knowledge engineer as in rule-base systems, but neural networks are trained from a large set of test data which represents the values for all independent and dependent variables (or the inputs and outcomes) of a system problem.

The neural nets used in this hybrid system are three layer networks with one layer of hidden nodes. Depending on the type of neural networks, their inputs can be either binary (Yes/No or On/Off) or analog (real numbers scaled between 0 and 1). To represent the sound and solar properties adequately, analog outputs were required.

4.2.3 Training and Testing Procedures

The networks were trained by a supervised learning algorithm using a back propagation error correction to adjust the weights associated with each node (Figure 4.4). During the training session, the weights of the network nodes are adjusted until the sum square error is minimized, somewhat similar to a regression analysis. In contrast to a regression analysis, the functional relationship between inputs and outputs does not have to be known and non-linearities are accommodated.

Neural networks can handle duplicate or contradictory training data and their learning process reduces the noise in the training set. When the trained neural network is later confronted with a set of inputs that did not exist in the original training set, the network will give a best estimate based on the general features it detected in the training set. Of course, the accuracy of such a prediction depends on the how many training examples were available and how accurate they were, as well as on the structure of the neural network.

O1: Sound transmission class (STC) value (effectiveness rating)

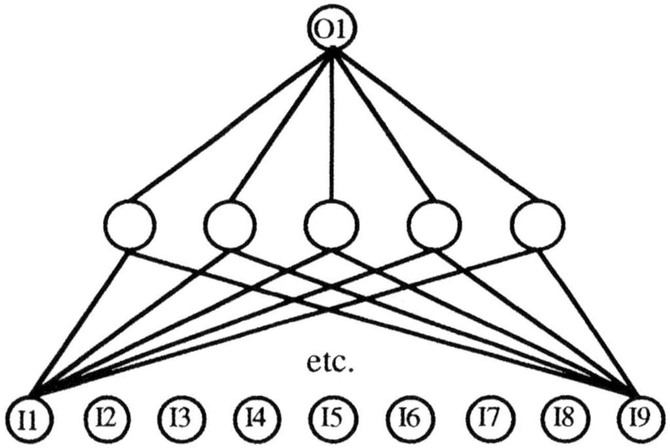

I1: Inside glazing or laminate glass lite or monolithic glass
I2: Inner lite vinyl PVB thickness (if inside lite laminated)
I3: Inner glass lite facing outside (if laminated inner pane)
I4: Extra vinyl layer in outer ply (if triple ply laminate)
I5: Third glass lite facing outside (if triple ply laminate)
I6: Construction internal air space (if insulated design)
I7: Outside glass lite facing inside (if laminated outer pane)
I8: Outer lite vinyl PVB thickness (if outside lite laminated)
I9: Outside glazing pane or laminated glass lite, facing outside

Figure 4.4 Neural network structure for sound control data

Table 4.1 Comparison of Interpolations to Training Sets for Sound Control Net

STC versus Layer Thickness	Neural Net	Multiple Linear Regression
Avg. Abs. Err. in STC Units	+/- 1.0	+/- 1.2
Avg. Abs. Rel. Error	2.7%	3.1%
R Squared Value	0.909	0.861
Data sets with Rel. Err. > 9%	4 of 94	7 of 94

STC Units correspond to decibels from transmission loss spectra per ASTM Standard E 90-87 and ASTM Standard E 413-87.

O1: Visible light transmittance (fraction)
O2: Solar optical transmittance (fraction)
O3: Solar optical shading coefficient (fraction)
O4: Design instantaneous heat gain (BTUH/sq.ft.)

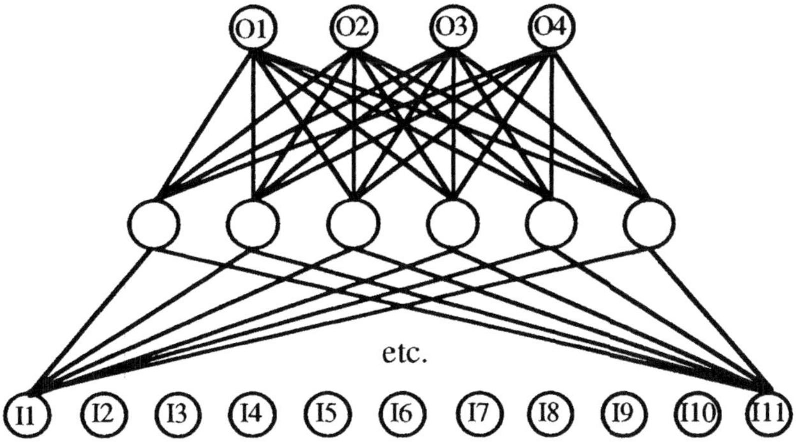

I1 - I 10: Binary representation of 10 standard PVB colors
I11: Total (clear glass) laminate thickness with 0.030 in. PVB

Figure 4.5 Neural network structure for solar control data

Table 4.2 Comparison of Interpolations to Training Sets for Solar Control Net

	Fractional			Solar Heat
Colors and Glass Thickness	Visible	Solar	Shading	Gain
Neural Net	Trans'e	Trans'e	Coeff't	(I-P)
Avg. Abs. Error in Units	0.006	0.008	0.007	5.05
Avg. Abs. Rel. Error	1.36%	1.42%	1.01%	2.96%
R Squared Value	0.999	0.998	0.997	0.961

4.2.4 Neural Networks for Plausible Estimates

The neural net for sound control was trained using primarily published data for sound transmission class (STC) for specimens where the transmission loss spectrum for the required 16 frequencies was available. Some proprietary data on laminates were also included. The neural net for sound control maps only the STC value to the thickness of the individual layers in the glass assembly, including each glass ply, PVB interlayer and, for insulated glass (IG units), air space thickness (see Figure 4.4). For 94 examples, the trained network [i=9,h=5,o=1] using five hidden nodes (h) to represent nine inputs (i) and one STC output (o), predicts non-linear outcomes better than multiple linear regression (Table 4.1).

Neural nets not only predict single outputs but can also represent non-linear systems with multiple outputs. For solar control, the neural net was trained using published laminate data for light and heat transmission (LSGA, 1985) and trade literature for clear glass. Proprietary data on 1/4-inch (3 mm) laminates were included. The neural net correlated four solar characteristics with ten standard PVB colors and the total glass thickness [i=11, h=6, o=4]. The ten colors were represented as a binary input array where all `off' or `zero' inputs represented clear solid glass; inputs were ordered by transmittance (see Figure 4.5). The solar neural net used six hidden nodes and represented the data well (see Table 4.2). The average relative error in predicting the training set data for solar control shading coefficient was about 1.0%, or 0.007 shading coefficient fractional units.

4.3 Operational use of LAM^{tm}

The system itself was a bridge to experts in the glazing industry. Communication and feedback with clients was enhanced by software deployment and discussion with users. The primary source of enhancements has come from hands-on keyboard work with prospective clients.

With laminator customers, LAM^{tm} has been demonstrated and used to initiate marketing dialogue regarding their DuPont supplier relationship. With a laminators trade association, LAM^{tm} was used to clarify existing standards on laminate design strength. LAM^{tm} has helped establish working relationships with industry glass experts among primary glass manufacturers needed to extend capability for laminate strength design. Documentation of the technical basis was published (Foss and Droste 1990).

Internally LAM^{tm} was used to supplement staff training on end-use performance characteristics of architectural laminates. For outside inquiries, LAM^{tm} has been used to respond to specific questions on particular glass applications including aquarium design, a special long term load case.

For glazing specifiers and architects, LAM^{tm} software was offered as a follow-up in conjunction with a major world-wide DuPont advertising campaign, called "IMAGINE - Light Space and Laminated Glass". As marketing communication software, LAM^{tm} was expected to benefit DuPont's architectural "Butacite" business. According to Ed Mahler, then manager of Decision Support Systems for DuPont Information Systems, "the LAM^{tm} concept was a splendid example of our market focus thrust of taking our expertise and putting it at the customer's design table."

The design calculations for wind load structural strength of glass are basically straightforward, but the various design constraints and code requirements make it a tedious and repetitive job. Even when a glass thickness appears acceptably strong, an array of other criteria must be serviceable, especially deflection. LAM^{tm} can simplify the window glass selection and its integration benefits ease-of-use. Integration has the advantage of producing a functional design across multiple criteria, while requiring less designer time.

4.3.1 The Software Development Process

LAM^{tm} has been deployed in two versions, English and metric units, and for a variety of different purposes. The English version in inch-pound units was distributed in the USA because it closely represents established standards for glass strength. The metric version was developed with user-definable strength factors, including K4 values as a function of thickness. This customization feature permits users to adapt the metric version to local methods and requirements.

This glass design software was ready to place in commercial service twelve months after its development was begun. The first prototype was demonstrated to potential users only two months after the start of the project and was based on strict representation of a standards manual (LSGA 1985).

Based on customer feedback, refinements were made over the next nine months, including modularization of the system's design which required rewriting all subroutines, and an extension of the knowledge content to a larger number of glass constructions and different design methods. The final month was required for approvals and preparation of distribution materials.

4.3.2 Relationship to Model Building Codes

For basic strength of annealed glass, LAMtm uses a statistical correlation of the maximum pressure-area product (K4) as it is defined in a sanctioned voluntary specification, ANSI/AAMA 101-88 (AAMA 1988). The USA model building codes differ to a degree from the maximum limits advised in LAMtm. For three nominal thicknesses, the K4 values represented by LAMtm are higher and therefore less conservative than USA model codes: 3/16" (5 mm), 1/4" (6 mm) and 5/16" (8 mm), but can be user-definable (see 4.3.1).

Table 4.3 Comparison of LAMtm glass strength to USA model building code maximums: inch-pound units version.

K4 = (pressure_PSF) x (Area_SqFt)

Nominal Glass Thickness (inch)	K4 Value Representing Building Codes (lb-force)	K4 Value Modeled In LAMtm Software (lb-force)	Equivalent LAMtm Design Probability (lites/1000)
Single Strength	267	245	12
Lami	320	287	13
1/8"	405	389	10
5/32"	610	565	12
3/16"	727	757	7
1/4"	1075	1161	6
5/16"	1544	1615	7
3/8"	2228	2102	11
1/2"	3252	3215	9
5/8"	4550	4496	9
3/4"	6200	5947	10
7/8"	7790 *	7565	9
1"	9800 *	NA	NA

* Certain USA building codes may not define glass thickness over 3/4"

However LAMtm also estimates the probability of glass fracture, where vertical glass is normally designed for a breakage probability of 8 lites per 1000. Tabulated is the LAMtm design probability where the maximum design load is equivalent to USA model building codes (see Talbe 4.3). For 3/16", 1/4" and

5/16" glass thicknesses, using a maximum LAM^{tm} design breakage probability of 6 lites per 1000 will insure compliance with codes. Designers must determine the suitability of each application use in practice, and comply with the provisions of the building codes (Vild, 1993). And consequences of glass breakage must be considered.

4.3.3 How the Software Is Used

The LAM^{tm} hybrid system is deployed as a stand-alone software package for DOS compatible personal computers, preferably equipped with a hard disk. The intended users are architects and glazing specifiers as well as the glass laminators.

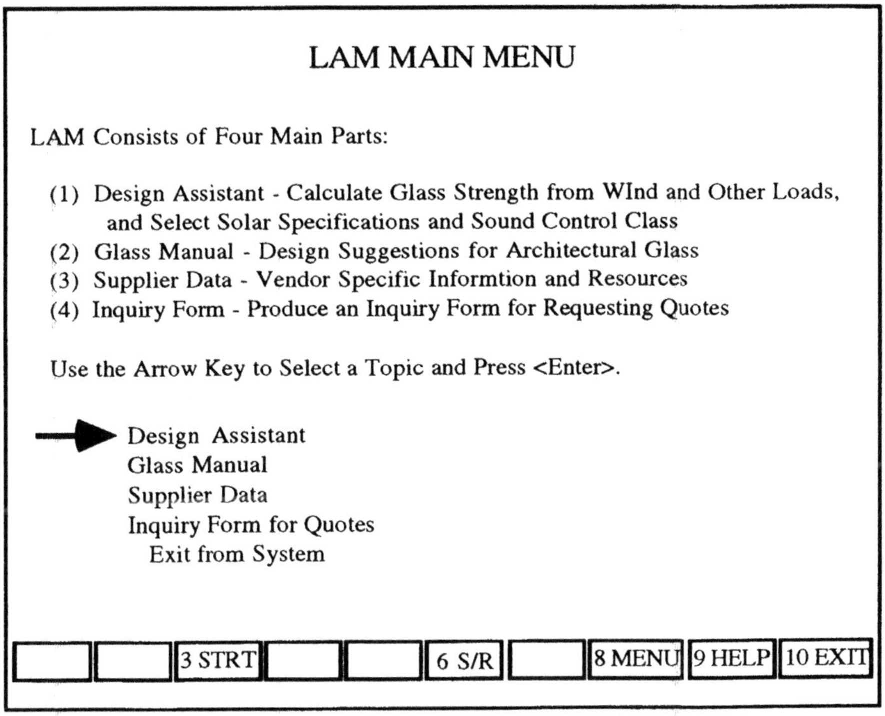

Figure 4.6 LAM^{tm} main menu

Upon start-up of the software, the user is asked to decide whether to design a particular glass unit or to browse through the text of the glazing manual (see Figure 4.6). If the design feature is chosen, the system carries out an interactive consultation which starts with the user selecting one of seven possible types of glass constructions for the design.

The user can now choose either a particular combination of glass thicknesses and determine whether the selected structure satisfies the loading conditions, or can input the external loading conditions to compute the minimum essential glass thickness. An algorithm from the Canadian method was used to estimate initial thickness (CGSB, 1989). Either way, after the glass dimensions are selected, the expert system gives the user a choice of determining the glass strength under wind loading, the sound transmission class of the construction or its solar properties.

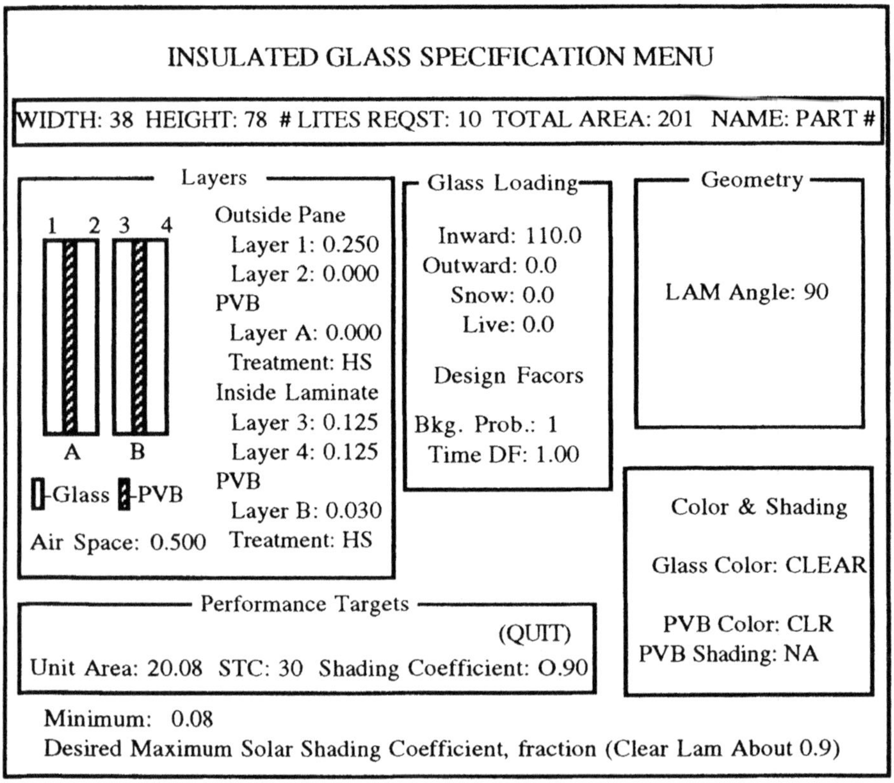

Figure 4.7 User data input screen

Glazing used for aquariums or port holes are designed similarly. Since only questions pertinent to the particular design are asked, and because it is easy to use, the glass design expert system will reduce the time required for the designer to determine the minimum glass thickness for the glazing application. The glass designer can also use the software to iteratively propose glazing designs and critique them for acceptability with respect to structural strength, sound control and solar control targets.

Once the selected glass construction has met all of the specifier's requirements, the underlying design information, the dimensions of the glass unit, and all of its determined properties can be stored in a data base record to be used later for review or as a template for a new design. In addition, the specifics of the selected glass construction can be combined with a file of preferred vendors to print out an inquiry form for quoting price and delivery of the glass design.

4.3.4 Discussion of Example

The sample calculation in Table 4.4 illustrates a case where strength prediction for the laminate deviates qreatly from a 1.0 laminate design type factor for a rigid shape. The 1/2 inch (12 mm) thick laminated specimen (Reznik and Minor, 1986, series "ADO") had a failure strength of 275 PSF (13.2 kPa) which if divided by the 2.5 glass design factor for a breakage probability of 0.008 gives an expected design strength of 110 PSF (5.30 kPa), used a target design input.

This experimentally expected design strength is below the design predictions based on both the straight line AAMA method and the CGSB approximation which uses a higher laminate type factor (122.4 and 121.7 PSF or 5.88 and 5.84 kPa, respectively). But, the "flexibility ratio" is below 150 and an appropriate laminate type factor is thus 0.75 for use with the newer non-linear failure prediction model in ASTM E 1300 (see section 4.4.1). This type factor would reduce the ASTM design strength for 1/2 inch thick monolithic from 135 PSF (6.48 kPa) to 101.25 PSF (94.87 kPa) for the corresponding laminate, expectedly more conservative than experimental. Thus the effect of shape on laminate strength seems to explain the relatively low observed breaking pressure for this case.

"Flexibility ratio", or width by thickness, is about 76. The 38 inch (965 mm) wide by 1/2 inch (12 mm) thick nominal laminate series measured 0.4750 (12.1 mm including PVB) having two glass plies averaging 0.2214 inch (5.62 mm) thick, with relatively thick 0.0322 inch (0.818 mm) PVB interlayer. This example illustrates certain shapes of thick, rigid laminates can deviate below the strength of the equivalent total monolithic glass thickness.

Table 4.4 Sample Calculation for Vertical 1/2" Thick Laminate

	Annealed Laminate, 4-Sided Support	
Width x Length x Thickness, inch	380 x 76 x 2-ply 1/4 inch	
Area, square feet	20.1	
Aspect Ratio, a/b	2.0	
Flexibility Ratio. b/t	76	
Experimental: Failure Strength, PSF	275.0 (Series ADO)	
Coefficient Variation (CV)	17%	
Results of Design Calculation from Different Standards:		
		Laminate Type Factor
Straight Line Design, AAMA 1988		
Strength, PSF	122.4	0.75
Probability	8 per 1000	
Area Limits?	Below	
Failure Prediction Model, ASTM E 1300		
Strength, PSF	101.25	0.75
Probability	8 per 1000	
Statistical Interpolation, CGSB 1989		
Strength, lb/ft2	121.7	1.00
Probability	8 per 1000	
Results of Expert System Predictions:		
Specified Design Load, PSF	110.0	
Automatic Thickness Estimate, inch	0.238	(each ply)
Thickness Classification, inch	0.313	(each ply)
Iteratively Selected Thickness, inch	0.250	(each ply)
Software Strength Estimate, PSF	120.2	
Strength Classification	acceptable	
Area Limits?	Below	
Estimated Breakage Probability	6 per 1000	
Estimated Deflection, inch	0.24	

4.4 Implications and Directions for New Technology

A significant benefit from creating LAM^{tm} was the communication processes leading to enhanced and expanded knowledge about the product functionalities themselves. This "new knowledge" can then be considered as potential improvements for the system knowledge base. All areas of the system scope have been affected.

The knowledge base is continually changing and developing, both internally as the content expands and in the external environment. A perspective on the value of the communication processes involved is important, as is the dialogue necessary to enhance the knowledge base. Communications are critical to developing integrated hybrid systems.

In addition to meeting marketing communication goals, important advances in the knowledge base represented have subsequently developed. Research effort persisted. And active participation in communication processes with the industry after LAM[tm] deployment were valuable.

This section updates key aspects of laminate technology and illustrates the related and ongoing communications involved. In each functional area represented, there are examples of enhancements to the knowledge base.

Table 4.5 Laminate Shape Definitions for ASTM E 1300

a = Long Side of Glass Plate
b = Short Side of Glass Plate
t = Thickness of Glass Plate
a/b = Aspect Ratio, AR
b/t = Width/Thickness (Flexibility) Ratio

Laminate Design Type Factor Decision Table
Interim Values for ASTM E 1300 Development

	Laminate Design Type Factor (LTF)	
	AR =< 2	AR > 2
b/t =< 150	0.75	0.75
b/t > 150	0.90	0.75

Basis: Rule 1. If AR > 2.0, Then LTF = 0.75
Rule 2. If b/t =< 150 and AR =< 2.0, Then LTF = 0.75
Rule 3. If b/t > 150 and AR =< 2.0, Then LTF = 0.90

4.4.1 Effect of Shape on Structural Strength of Laminates

Structural strength of glass laminates has progressed beyond representation as a single, simple factor relative to monolithic annealed glass. Mathematical modeling of stresses in laminated glass in four-sided support have indicated that there are combinations of shape or aspect ratio (a/b) and flexibility or stiffness (b/t) where laminated glass is probably stronger than equivalent monolithic annealed plate, perhaps due to membrane stresses (see Table 4.5). These resulted in new rule-based approximations to the effect of shape on laminate strength (Vallabhan *et al.*, 1991; Conlisk, 1992).

4.4.2 Solar and Thermal Control As Key Glazing Function

Solar control with laminated glass has progressed well beyond representation of tinted PVB interlayer colors originally in LAM^{tm}. An entire mathematical model, WINDOW 4.0, for calculation of solar and thermal performance properties of glazing assemblies has been developed by Lawrence Berkeley Laboratory, with Department of Energy support. Measurement and estimation of relevant laminated glass properties is a continuing challenge. And spectral selectivity, with high ratios of transmitted daylight in relation to solar heat is particularly advantageous as single laminated glazing in southern climates.

To develop capability beyond LAM^{tm}, active participation on representing "Butacite" tinted PVB laminates in WINDOW 4.0 was engaged. This led to an interesting inquiry from the Florida Solar Energy Center regarding wise glazing choices for rebuilding after Hurricane Andrew. Because laminates can have the missile penetration resistance important to building integrity in hurricanes, the inquiry was taken very seriously and a new class of "Low-IR" solar laminates was conceived:

"The second approach is to laminate a spectrally absorbing tinted glass to a clear glazing with a pyrolytic [exposed interior] low-E coating. These coatings have only modest emittance value. This would be essentially the same as the [monolithic] approach but the laminate would conduct less of the heat absorbed by the tinted outer glazing and this approach might offer greater choice in the inner and outer panes. It would also offer improved shatter and possibly penetration resistance. Improved noise control is another benefit because the typical laminated 2-ply construction reduces sound transmission over a wide frequency range, depending upon glass and interlayer thickness. The polyvinyl butyrate (PVB) used in laminated architectural glass filters out some of the UV radiation incident on the glass, its protection varying with type and thickness of the PVB layer. UV protection can be important due to its contribution to color fading of fabrics and other interior furnishings and material damage to some plastics and other materials." (McCluney, 1993)

Table 4.6 Sound Transmission Characteristics of Single Glazed Laminates Comparison of Measured "STC" to Neural Net Model

RAL Test Report	--Glazing Layer Thickness-- Exterior (inch)	PVB (mil)	Interior (inch)	Neural Net STC	Test Report STC
	0.250	0	0	27.5	27 to 32 expected
TL89-364	0.125	15	0.125	32.0	34
TL89-363	0.125	30	0.125	33.1	34
TL89-362	0.125	45	0.125	34.0	34
TL89-365	0.125	60	0.125	34.9	35
	0.378	0	0	29.8	28 to 33 expected
by others	0.188	15	0.188	34.8	36
by others	0.188	30	0.188	35.5	36
estimate	0.188	45	0.188	36.2	NA
estimate	0.188	60	0.188	36.8	NA
	0.500	0	0	31.6	31 to 36 expected
TL89-367	0.250	15	0.250	36.5	37
TL89-371	0.250	30	0.250	37.1	37
TL89-372	0.250	45	0.250	37.6	38
TL89-373	0.250	60	0.250	38.0	38

4.4.3 Sound Transmission Measurement and Modeling Refinements

Sound attenuation is measured as a function of frequency. The resulting transmission loss spectrum is classified as a single value, such as the Sound Transmission Class (STC) used in LAMtm. But representation of sound control has progressed beyond the single measure of STC. An Outdoor Indoor Transmission Class (OITC) is intended to quantify lower frequency road noise, per ASTM E 1332. And in Europe, another classification is defined as the weighted sound reduction index (Rw), per ISO 713/3. The "mass law" theoretically defines the upper acoustic spectral performance limit, improving with mass per unit area. The heavier the glass the better. Two properties cause relative performance dips: low frequency resonance characteristics related to stiffness and mass properties; and high frequency coincidence characteristics affected by interlayer damping properties. Neural net STC model can be compared to laminate acoustic measurements from Riverbank Acoustical Laboratory (RAL).

Table 4.6 illustrates use of the neural net in LAM^{tm} free standing, without the LAM^{tm} user interface, to examine design cases. LAM^{tm} was used in responding to a wide variety of client inquiries on acoustics. And such inquiries stimulated a variety of valuable technical communications on enhancements to acoustical control and representation.

4.4.4 Penetration Resistance to Windborne Missile Impact

Windborne debris is much more likely to break glass than is wind pressure (Norville, 1992). Penetration resistance of laminated and other multi-layer glazing systems is an important functionality omitted from LAM^{tm} because the relationship of glazing layer construction to comprehensive measures of impact resistance is simply not available. Impact methods developed around the world vary, and tend to be specific "pass/fail" tests for particular applied kinetic energy, sample areas and support.

Importantly, building code bodies in southern USA are considering a large body, timber missile impact test for glazing in hurricane prone coastal regions as a result of Hurricane Andrew damage in south Florida. A precedent in Australia exists as a result of 1974 cyclone Tracy damage to Darwin (TR-440, 1988).

A penetration resistance test method is under consideration within the ASTM consensus process (Dolenga, 1993; Beers, 1993; Minor, 1993). This illustrates the communication benefits resulting from exploring the limitations of the LAM^{tm} design system and pursuit of enhancements to its knowledge base.

4.5 Conclusions

This software illustrates how the design of a building component, the glazing, for different functional capabilities can be simplified for the user. The integration of design knowledge within an expert system has the important advantage of producing a design that is both functional and consistent across various design criteria while requiring less time from the designer.

Neural net representation appears to be practical for estimating design data when a sizable set of test data exists that correlates output predictions accurately with input parameters. The average error between the training set data and the neural net predictions can be controlled to a level useful for design, both by properly structuring the network, and by selecting an appropriate number of hidden nodes for the amount of data available.

Controversy exists concerning the design factor approach for glass laminates relative to monolithic glass. While the newer design methods are progressing toward functional relationships describing laminate design, USA model building codes and ANSI/AAMA 101-88 presently have established a laminate design factor of 0.75 times the strength of equivalent total glass thickness (AAMA 1988). Additional mathematical modeling and tests on laminate structural strength are leading toward a new USA consensus design practice for selecting glass thickness to resist wind loads, ASTM E 1300, using rules to approximate laminate shape effects.

Hybrid systems as a means to integrating broad areas of knowledge are, in some sense, an embodiment of a kind of man-machine symbiosis. The system representation can positively influence the knowledge represented. And the interaction among the people involved in the system's knowledge acquisition and representation enhances their communication processes.

Acknowledgments

LAM[tm] glazing design software was developed to help both designers and DuPont staff better understand the benefits and limitations of architectural laminates. By funding and supporting this project, the DuPont Company shares expertise in artificial intelligence with "Butacte" laminator customers. "Butacite" and "SentryGlas" are DuPont trademarks for proprietary polyvinyl butyral sheeting compositions, "ChromaFusion" and ContraVision" are Cesar Color Incororated trademarks. Sources for some of the data and additional laminate lateral pressure tests include bulletin H-45102, "Laminated Glass with DuPont Butacite"; suitability for use must be determined for each application. Features and benefits of architectural laminated glass are from the media kit of the DuPont Benedictus Award, presented during the June 1993 American Institute of Architects (AIA) convention for innovative use of laminated glass in London's third airport at Stamsted.

Chapter 5

Hybrid System Approach to Nuclear Plant Monitoring

5.1 Introduction

This chapter is based upon a final report prepared by Charles River Analytics Inc., Cambridge, MA for the U.S. Department of Energy, Germantown, MD under Contract No. DE-FG02-91-ER81178. It describes the results of contract efforts to develop a nuclear plant monitoring system based on a hybrid neural network knowledge base approach. The program was directed by Mr. James M. Mazzu of Charles River Analytics Inc. with Dr. Alper K. Caglayan as the program manager. Mr. Paul G. Gonsalves contributed to the project.

The major objective of nuclear plant monitoring systems is to enable the operator to make correct decisions. The usual difficulties encountered by nuclear plant operators are incorrect assumptions of plant status, insufficient information for fault diagnosis, and prediction of scenarios resulting from various courses of action. To alleviate these difficulties, nonalgorithmic approaches based on neural networks can clearly be of help by replacing and/or augmenting classical analytic measurement validation techniques. Similarly, expert system rule bases can augment the classical hypothesis testing techniques by incorporating operational constraints. Here, we use the complementary strengths of neural networks and knowledge-based expert systems to create a nuclear plant sensor monitoring system that can outperform either method alone.

In this study, we developed a hybrid neural network/knowledge base approach to intelligent sensor monitoring in nuclear power plant applications, developed a hybrid neural network/knowledge base plant monitoring architecture, implemented a prototype plant monitoring system using our neural network/expert system shell NueX on the Macintosh platform, and showed the feasibility of our approach by detecting failures in neuron detectors using in-core neuron flux samples provided by Yankee Atomic.

Potential benefits of hybrid nuclear monitoring system are enhanced operator efficiency and performance by providing intelligent operator recommendations, enhanced productivity of the plant operations by detecting off-normal operating states before they drastically affect plant operating performance, increased plant safety by detecting off-normal operating states before they degrade the safety of plant operations, reduced plant maintenance costs by identifying subsystem and component failures before they lead to subsequent failures of related components, and improved operator training in understanding the diagnostic process.

5.1.1 Background

The major objective of nuclear plant monitoring systems is to enable the operator to make correct decisions. The usual difficulties encountered by nuclear plant operators are incorrect assumptions of plant status, insufficient information for fault diagnosis, and prediction of scenarios resulting from various courses of action. The problems of plant mode status determination, information management for fault diagnosis, and prediction of results of controller actions fall into the domain of knowledge-based situation assessment and decision aiding in the domain of artificial intelligence. Here, we use a hybrid neural network and expert system approach to develop an intelligent monitoring system for nuclear plant applications.

Clearly, the starting point of a hybrid approach should be the current state of the art in nuclear monitoring systems. Moreover, the emphasis of an intelligent monitoring system should be the shortcomings of current systems, for instance, hardware redundancy-based measurement validation has been fairly successful. In contrast, analytic redundancy-based measurement validation has not been universally accepted in the industry due mostly to the limitations of analytic plant models. Here, nonalgorithmic approaches based on neural networks can clearly be of help by replacing and/or augmenting analytic measurement validation techniques. Neural networks are used to develop plant models to be used for fault diagnosis instead of the classical parameter identification techniques such as parity space checks, multiple model tests, etc. Similarly, expert system rule bases augment the classical hypothesis testing techniques by incorporating operational constraints.

Recently, artificial neural networks (ANNs) have been applied to a variety of pattern recognition problems regarding fiber optic sensors (Mazzu, Caglayan, Allen, 1992), control reconfiguration (Caglayan, Allen, and Rahnamai, 1989), space vehicle guidance (Caglayan and Allen, 1990), multiple target recognition (Gonsalves and Caglayan, 1992), and remote sensing (Mazzu, Snorrason, and Caglayan, 1992). The advantages of ANN include the ability to classify patterns that vary in an unknown manner, recognize patterns within noise, and recall patterns even if some processing units fail. However, ANNs fall short where knowledge-based expert systems (KBs) excel; such as handling formal logic, heuristics, and domain knowledge. In order to benefit from the capabilities of each method, sensor monitoring systems are being created within hybrid ANN/KB environments (Mazzu, Caglayan, and Jonas, 1992) and therefore have the potential for unsupervised adaptation. This is critical for providing intelligent adaptation to unexpected conditions and varying domain applications.

The integration of ANNs and KBs is an ideal step in the development of intelligent systems. In general, the two methods complement each other such that ANNs provide soft constraints, while expert systems allow hard constraints (Glover and Rao, 1990). Specifically, ANNs perform nonlinear functions, pattern recognition capabilities, fault tolerance and parallel processing; while expert systems involve language processing, formal logic and rule interpretation. Here, we exploit the complementary strengths of neural networks and knowledge-based expert systems to create a hybrid intelligent sensor monitoring system for nuclear power plants that can outperform either method alone.

5.1.2 Summary

In our work, we developed neural network and hybrid neural network/knowledge-based strategies for intelligent sensor monitoring capabilities within nuclear power plant applications. The neural network focus involves determining plant operating states including normal and off-normal conditions, and investigating the temporal signatures of the various sensor measurements. The hybrid strategies focus on integrating the ANN results with expert assessment and prediction knowledge in order to provide the plant operator with intelligent recommendations. We investigated neural network and hybrid neural network/knowledge-based strategies for developing intelligent nuclear plant monitoring systems. In particular, answers to the following questions are sought:

- How can the unique capabilities of neural networks be maximized to provide sensor pattern recognition solutions for nuclear plant monitoring?
- Which neural network paradigms will be most appropriate for process and structural sensing functions? Will time dependent paradigms be necessary?
- What are the most promising neural network internal architectures, including number of layers, connection paths, bias values, and activation functions?
- In what ways will the integration of neural networks with expert systems provide a significant advance for nuclear plant sensor monitoring?
- What hybrid neural network/knowledge-based strategies can be developed for accurately recognizing and interpreting process sensor indications?
- How can the on-line adaptation capabilities of a hybrid system be utilized within nuclear plant sensing applications?

5.2 Formulation of the Problem

5.2.1 Industry Contacts

Throughout this effort, nuclear industry contacts have been established to assure successful acquisition of sensor data and to provide for continuous consulting. The established contacts include Richard Cacciapouti and Paul Theriault from Yankee Atomic Electric Company, and Paul Gurney from Seabrook Nuclear Power Station. Mr. Cacciapouti and Mr. Theriault have provided in-core neutron flux measurements from the Maine Yankee plant and Mr. Gurney has provided thermal-hydraulic data from the Seabrook Station.

5.2.2 Nuclear Plant Process Variables

The Seabrook Station thermal-hydraulic data consists of continuous hourly recordings for the months of June, July, October, and November of 1991. The measurements are obtained from Seabrook's process computer which has available approximately 4600 measurement variables from which to select.

The Maine Yankee in-core instrumentation data consists of weekly neutron flux measurements for a continuous 52 week period within 1991. Table 5.1 lists the variables represented within the Maine Yankee in-core data. Figures 5.1 through 5.5 illustrate the flux (for a sample detector), power, rated flow, inlet temperature, and RCS pressure, respectively. As illustrated, the drops in flux at weeks 15 and 51 (Figure 5.1) are directly correlated with the drops in power (Figure 5.2). This fact is utilized within the ANN architectures as described in section 5.4.3.

Table 5.1 Maine Yankee In-Core Variables

DESCRIPTION	UNITS
Neutron Flux	mV
Power	MWT
Rated Flow	%
Inlet Temperature	F
RCS Pressure	psia
Detector Location	Rod/Level

The flux detector locations are identified by position within the core and by the corresponding level, each core position consisting of four possible detector levels. Due to a core tilt of less than 0.3%, detector measurements from symmetrically located core positions are considered to be correlated, and therefore provide a comparison measure from which off-normal operating states are determined. Represented within the Maine Yankee data set are detectors (delayed response Rhodium Combination (RhC) detectors) correlated with one, two, and three other detector locations; for a particular detector location, its four detector levels are not necessarily considered to be correlated.

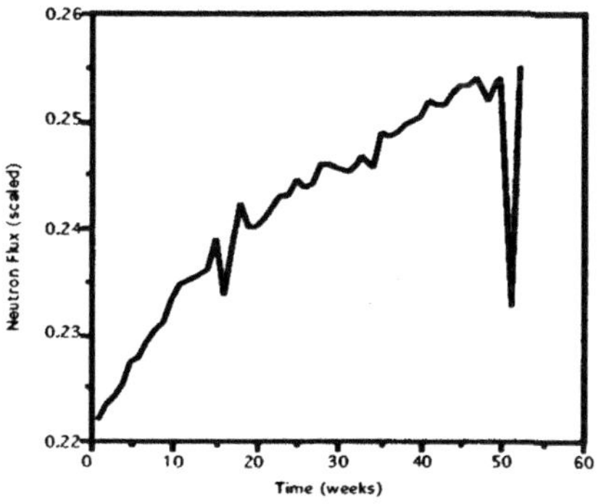

Figure 5.1 Sample Detector Neutron Flux

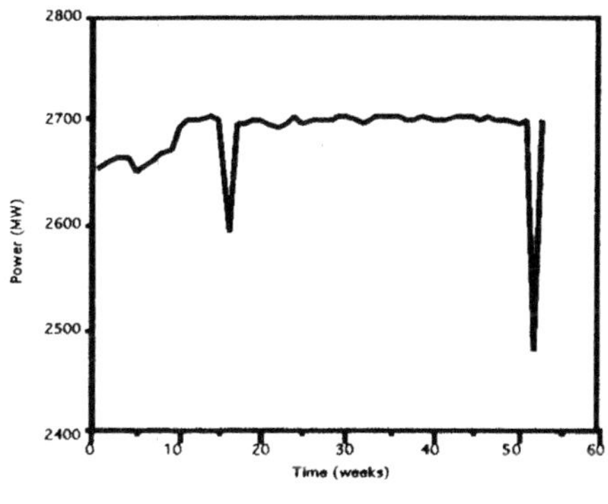

Figure 5.2 Power

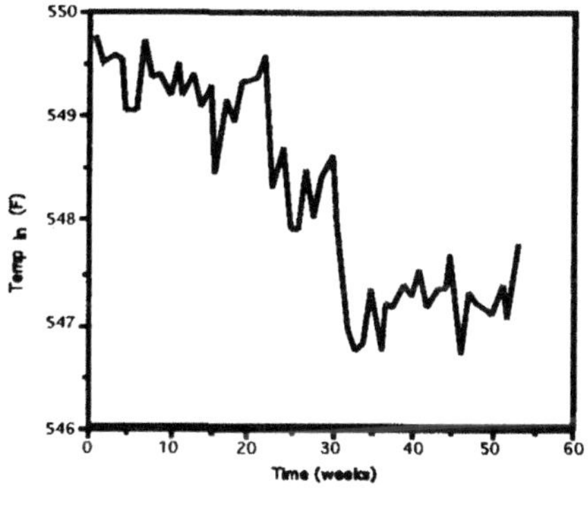

Figure 5.3 Inlet Temperature

5.2.3 Off-Normal Operating States for Neutron Flux Detectors

The off-normal operating states that may be present within the neutron flux data set consist of

- Detector Failed Hard
- Detector Failed Gradually

- Detector Failed Soft
- Flux Oscillations between Levels
- Flux Oscillations around the Core

Figures 5.6 through 5.9 illustrate the off-normal detector failures classified as Hard, Gradual, Soft (failed low), and Soft (failed high), respectively. Each figure consists of a neutron flux comparison between two correlated detector pairs, for a given level. The Hard failure state consists of very large temporal and spatial differences between correlated detector measurements.

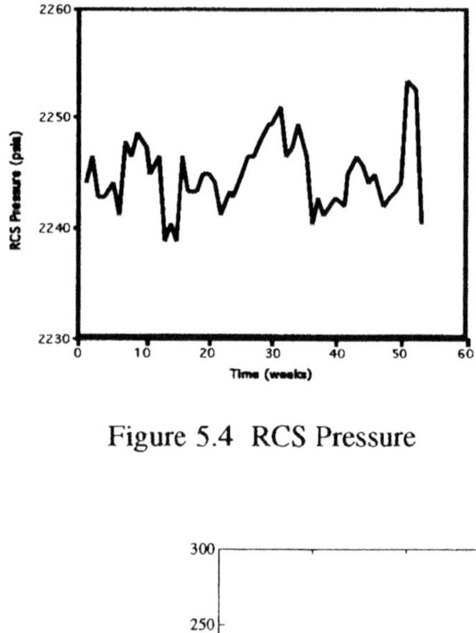

Figure 5.4 RCS Pressure

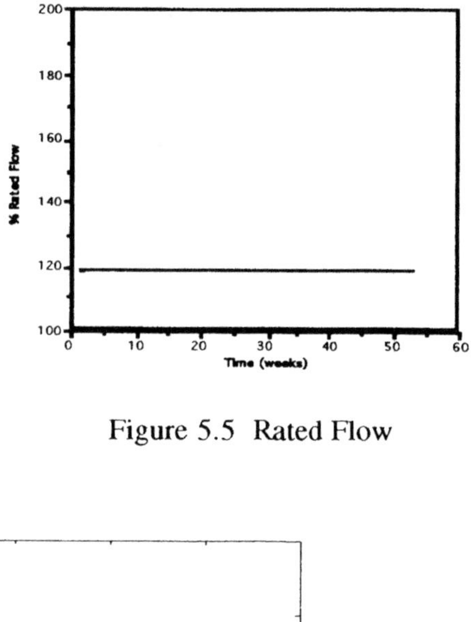

Figure 5.5 Rated Flow

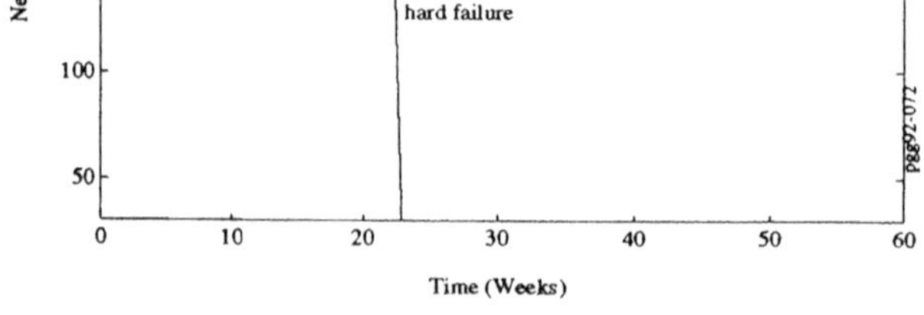

Figure 5.6 Detector Failed Hard

The Gradual failure state consists of a temporally expanding spatial difference, yet no detectable temporal change from one time step to another. One variation of a Soft failure state consists of the detector failing low with a detectable temporal change, then maintaining the failed state with a significant spatial difference between correlated detector measurements. Another Soft failure state (in this case failing high) is one in which there is a significant temporal and spatial change that only lasts for one time step.

As shown, it is very easy to distinguish the Hard failure state from the other more subtle Soft and Gradual detector failure states. For this reason, a hybrid neural network and expert system approach will be incorporated to ensure that all off-normal states are detected accurately. Neural networks will be able to identify Soft and Gradual failure states most effectively while the Hard failure

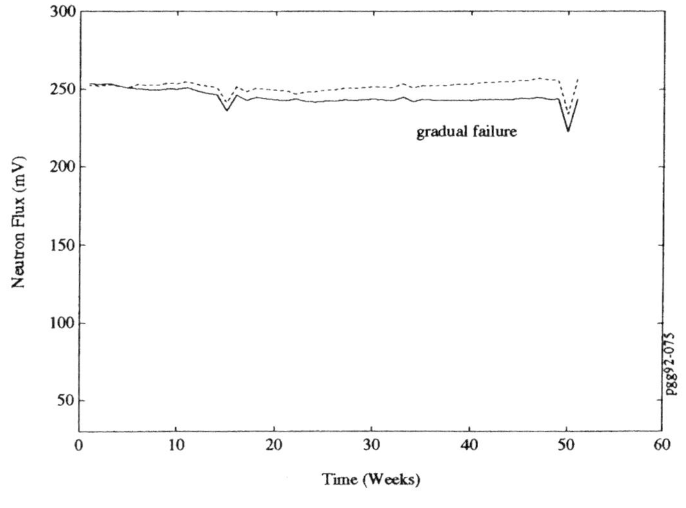

Figure 5.7 Detector Failed Gradually

states are monitored and identified by a rule-based expert system. In this manner, the Hard failure state information will not drown out the Soft/Gradual failure states during neural network training. Therefore, in the integrated hybrid nuclear monitoring system, the complementary strengths of neural networks and expert systems will work together to provide accurate off-normal state detection.

Here we have investigated the various off-normal failed detector states. A more frequent sampling of neutron flux is necessary to address the off-normal state of flux oscillations, both within levels and around the core. This further investigation is intended as part of the Phase II effort.

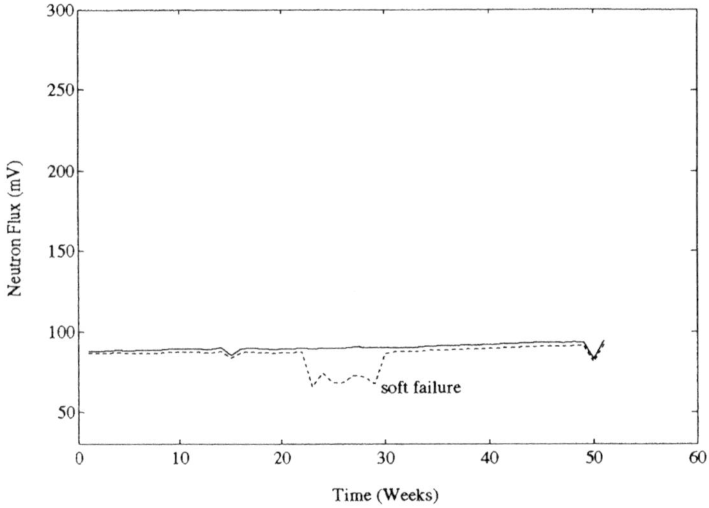

Figure 5.8 Detector Failed Soft (failed low)

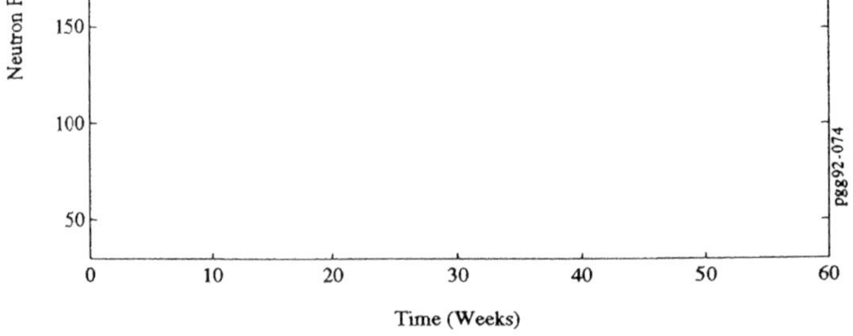

Figure 5.9 Detector Failed Soft (failed high)

5.3 Identification of Neural Network Strategies

5.3.1 Reduction of Nuclear Process Data for ANN Training

For the nuclear reactor core sensor failure detection problem there is an abundance of data for use in training. This data needs to be reduced not only for reducing the training time, but also for finding conditions such as conflicting data that inhibit learning. Here, we present a unified vector space based approach to training data selection geared at finding conflicting data, redundant data, and data clusters. The relevance of this approach differs with the various ANN paradigms. For instance, ART networks can find the clustered data classes directly. However, the current effort incorporates only the backpropagation paradigm, therefore investigation of data clusters are necessary.

The feedforward multi-layered network is a function mapping the n-dimensional vector space $[0,1]^n$ into the m-dimensional vector space $[0,1]^m$.

$$f: X \to Y \quad \text{with } y=f(x) \quad \text{for } x \in X,\ y \in Y \text{ and} \tag{5.1}$$

$$X = [0,1] \times [0,1] \times \ldots \times [0,1] \quad \text{n-times} \tag{5.2}$$

$$Y = [0,1] \times [0,1] \times \ldots \times [0,1] \quad \text{m-times} \tag{5.3}$$

where [0,1] is the closed interval between 0 and 1 on the real line, n is the number of network inputs, and m is the number of network outputs. In order to analyze the training data for conflicting and redundant input/output pairs, we need a distance measure between two input vectors, between two output vectors, and between two input/output pair vectors.

On either $[0, 1]^n$ or $[0, 1]^m$, there are a number of distance metrics derived from the following norms:

$$d_a(x(k), x(j)) = |x_1(k) - x_1(j)| + |x_2(k) - x_2(j)| + \ldots + |x_n(k) - x_n(j)| \tag{5.4}$$

$$d_m(x(k), x(j)) = \max_i |x_i(k) - x_i(j)| \tag{5.5}$$

$$d_2(x(k),x(j)) = \left((x_1(k) - x_1(j))^2 + \ldots + (y_n(k) - x_n(j))^2\right)^{1/2} \tag{5.6}$$

where $x_i(k)$ is the i'th input in the k'th input/output pair. In the following, we use a generic distance d(x, y), although we have found the max-norm induced distance metric $d_m(x, y)$ to be the most convenient in our studies.

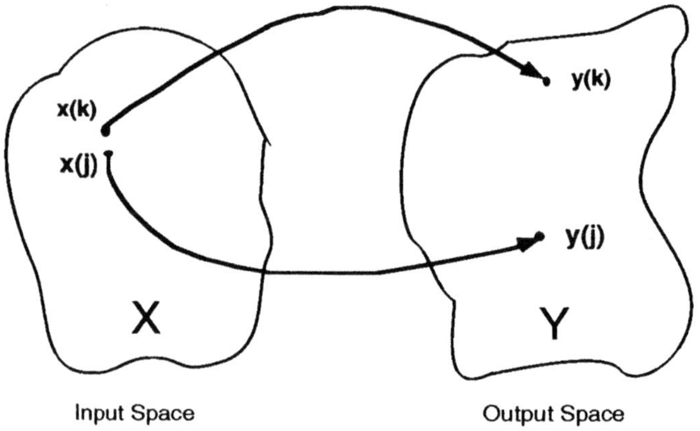

Figure 5.10 Conflicting Data

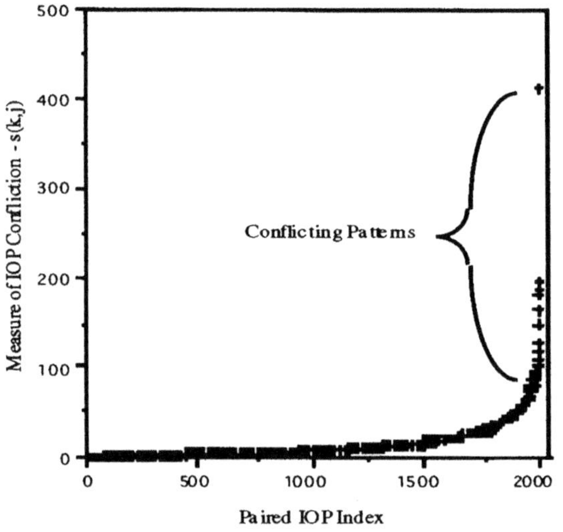

Figure 5.11 Sample Conflicting Training Data Analysis

5.3.2 Conflicting Data

Conflicting data are two or more input/output pairs where two similar inputs are mapped into dissimilar outputs (see figure 5.10). These input/output pairs violate the definition of a function where one input is mapped to one and only one output, thus making the training of a network impossible.

In analyzing the training data to find conflicting input/output pairs, the following function:

$$s(k,j) = \frac{d(y(k), y(j))}{d(x(k), x(j))} \tag{5.7}$$

gives a measure for the degree of conflict between the input/output pair $\{x(k), y(k)\}$ and $\{x(j), y(j)\}$. The higher the value of s, the higher the degree of conflict between the two input/output pair vectors.

This procedure is developed for determining the conflicting data within the reactor neutron flux ANN training data. Figure 5.11 plots the function s(k,j), which measures the degree of conflict for the sample training database. As shown, most of the points lie below an "s" value of 100. However, several data points exceed 100, with one point in the 400 to 500 range. The corresponding input/output pairs associated with these high values of s (or high level of confliction) are prime candidates for deletion from the training database.

5.3.3 Redundant Data

Redundant data are two or more input/output pairs where two nearly identical inputs are mapped into similar outputs. While this condition will not prohibit learning, it would adversely affect training time.

In order to analyze the training data to find redundant input/output pairs, we need a distance measure on the Cartesian product X x Y, which is the n + m dimensional vector space $[0,1]^{n+m}$. We can define distance measures on X x Y in terms of the distance measures on X and Y by:

$$d_a(z(k), z(j)) = d_x(x(k), x(j)) + d_y(y(k), y(j)) \tag{5.8}$$

$$d_m(z(k), z(j)) = \max_{x,y} \{d_x(x(k), x(j)), d_y(x(k), x(j))\} \tag{5.9}$$

$$d_2(z(k), z(j)) = \left(d_x^2(x(k), x(j) + d_y^2(x(k), y(j))\right)^{1/2} \tag{5.10}$$

where z(k) and z(j) are the k'th and j'th input output pair vectors defined by:

$$z(k) = \{x(k), y(k)\} \tag{5.11}$$

$$z(j) = \{x(j), y(j)\} \tag{5.12}$$

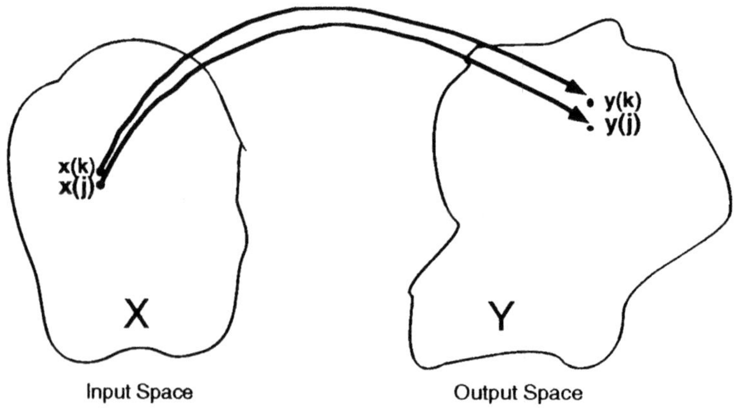

Figure 5.12 Redundant Data

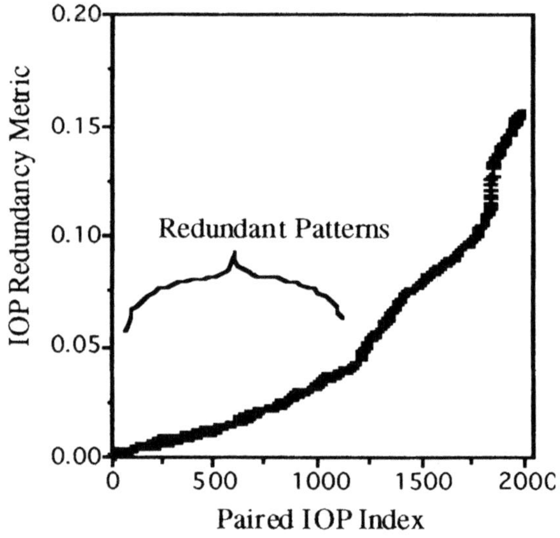

Figure 5.13 Sample Redundant Training Data Analysis

We have found the max-norm induced metric $d_m(z(k), z(j))$ to be the most convenient. Any of these metrics can be used to find almost identical data (where the distance between the input/output pair vectors would be nearly zero).

Determination of redundant data is developed for evaluating the flux detector ANN training data, and is illustrated by the plot shown in figure 5.13. A metric for data redundancy is plotted versus the associated input/output pair (IOP) index. The relevant feature to note is that the smaller the value for the redundancy metric, the more redundant those input/output pairs are. Therefore, from the figure, the pairs having a metric value less than 0.05 are candidates for elimination due to redundancy.

5.3.4 Clustered Data

Clustered data are the subsets of the training data, where the input/output pairs in each cluster subset are redundant. Data clusters can be found using the notion of formalized equality sets in (Lorczak, Caglayan and Eckhardt, 1989) with a distance metric on X and Y. According to this approach, a subset S of the input/output pairs is a formalized equality set if

$$d(z(k), z(j)) \leq \varepsilon \tag{5.13}$$

for any two input/output pairs z(k) and z(j) in S. Here, ε is a small number dictating the desired granularity. This clustering analysis can be performed to select a median input/output pair from each cluster class.

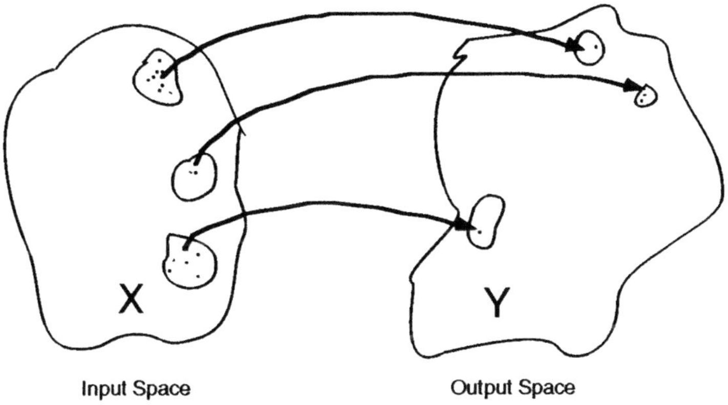

Figure 5.14 Clustered Data

A sample clustering of training data inputs is shown in figure 5.15 for the reactor core soft failure detection ANN inputs. The sample shown is for a network with two inputs, two outputs, and 140 training input/output pairs. The first 70 pairs produce the same output levels. If we concentrate on these first 70 pairs, we see approximately three data clusters: a cluster with input levels around 0.25, another cluster at 0.4, and one with input levels of about 0.5. The training database could be reduced by selecting the boundary input/output pairs from these clusters and eliminating the rest of the pairs within the cluster. A similar approach is undertaken for pairs associated with different output levels.

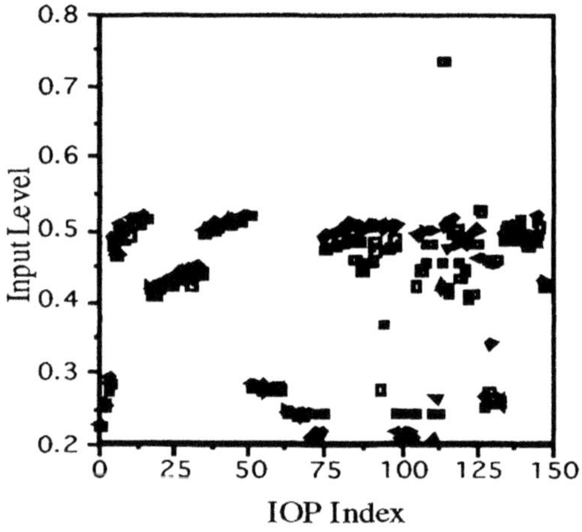

Figure 5.15 Sample Clustered Training Data Analysis

5.4 Neural Network Development

5.4.1 State Detection ANN with Temporal Feedback

The neural network development begins with the backpropagation network architecture shown in Figure 5.16. This architecture, and all subsequent architectures considered, is implemented with the in-house developed *NueX* hybrid environment. The ANN input nodes are represented as clear circular icons, the first hidden layer of nodes has the number "1" within the node icon, and the output layer nodes are shaded in gray. Information is processed from the input nodes, along the paths indicated with arrows, through the hidden layer nodes to the output layer nodes.

The output nodes represent the state of the correlated detectors (A and B) which are currently being processed by the network. If the output node's activation value is high (near 0.9), then the ANN has recognized a failed state for that corresponding detector. A low output value (near 0.1) indicates that the ANN has recognized a normally operating detector. The inputs to this network consist of each detector's neutron flux measurement for the current and previous data sample (the flux measurements were sampled on a weekly basis), as

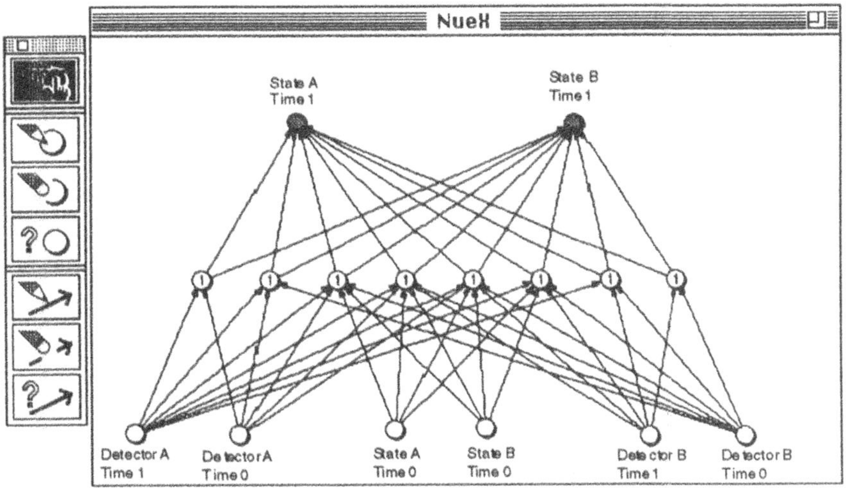

Figure 5.16 State Detection ANN with Temporal Feedback

indicated by “Time 1” and “Time 0”. Additionally, the assessed state of each detector for the previous data sample is fed back as input for the current data sample. These feedback input nodes are identified as “State A, Time 0” and “State B, Time 0”.

The training of this network consists of implementing the backpropagation paradigm with a set of 140 input/output patterns which represented normal, soft failure, return from soft failure, and gradual failure detector states. After approximately 2000 training cycles, this network successfully converged to an RMS training error of less than .06.

Upon evaluating this ANN with a correlated detector pair's yearly data samples (of which the network was not trained), the open-loop results (feeding back known detector states as inputs) indicate successful soft failure detection. However, the closed-loop testing (feeding back the actual ANN output states as inputs) reveals that the network must identify a failure state the first time it appears in the temporal data sample, or it will be relying on false results being fed back as input. In this case, the network cannot recover from the first missed detection and never picks up the failure state in successive weekly data samples.

5.4.2 State Detection ANN with No Conflicting/No Redundant Training Data

Due to the closed-loop difficulties encountered with temporal feedback of

ANN detection states, the temporal feedback nodes are removed from the network inputs. This leaves the temporal window of one time step (one week intervals) for each of the correlated detector pairs as the four ANN inputs. The network output nodes remain the state of each detector for the current data sample being investigated by the network.

When training with the original set of 140 input/output patterns, this ANN architecture has considerable difficulty converging to an RMS error of less than 0.4. For this reason, the strategies for determining *conflicting* and *redundant* training data are developed, as outlined in section 5.3. Upon applying these strategies to the neutron flux training data, it is evident that conflicting patterns and redundant patterns do exist within the training set (figures 5.17 and 5.18).

After removing the conflicting and redundant training patterns, there are a total of 70 input/output patterns remaining. With this reduced training data, the network without feedback converged to an RMS error of approximately 0.29, which is still too high for useful determination of failure states.

5.4.3 State Detection ANN with Clustered Data Boundaries

In order to investigate and further enhance the training performance, the strategy of data *clusters* is developed, as outlined in section 5.3. Upon plotting the training patterns inputs, the presence of clusters are apparent as illustrated in Figure 5.19. Using this cluster plot as a guide, the minimum number of cluster boundaries are determined.

From Figure 5.19, it can be seen that there are two major flux ranges, one around a scaled value of 0.5 (high range) and the other around 0.25 (low range). This table lists the cluster boundaries for the high flux range, and is mirrored identically for the low flux range. For the temporal window of two consecutive data samples (time 0 to time 1), there are four possible state changes: normal to normal, normal to failed, failed to failed, and failed to normal. Each of the fail states also has two possible directions, failed high and failed low (as illustrated in section 2.3). Furthermore, these failure states can occur on either of the correlated detectors (A or B) for each detector pair. For the "normal to normal" state changes, it is possible to have both large and small power drops (and power returns) which must also be accounted for so as to minimize false alarms. For the Gradual failure cases, the ANN is trained to output failed states for both of the correlated detectors. With a temporal window of one time step, this Gradual case is indistinguishable from a continued Soft failure state (shown previously in Figure 5.8). Therefore, further analysis will be conducted by a knowledge base within the complete hybrid monitoring system.

One training pattern is assembled for each of these minimum cluster boundaries, the total of which amounts to 34 training patterns. The training

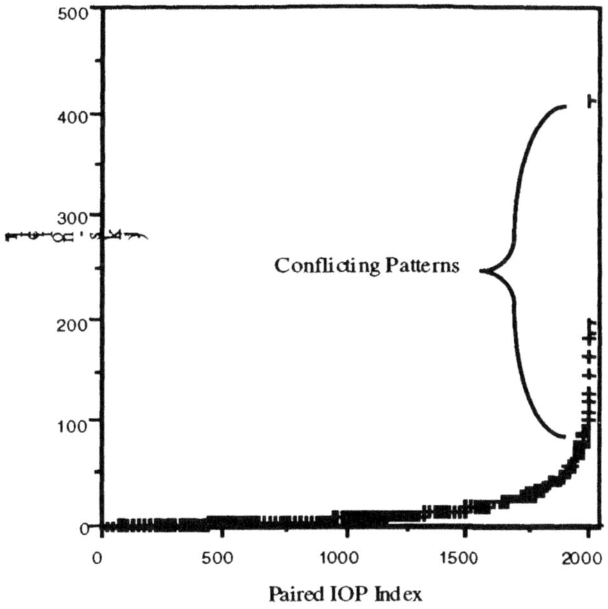

Figure 5.17 Sample Conflicting Training Data Analysis

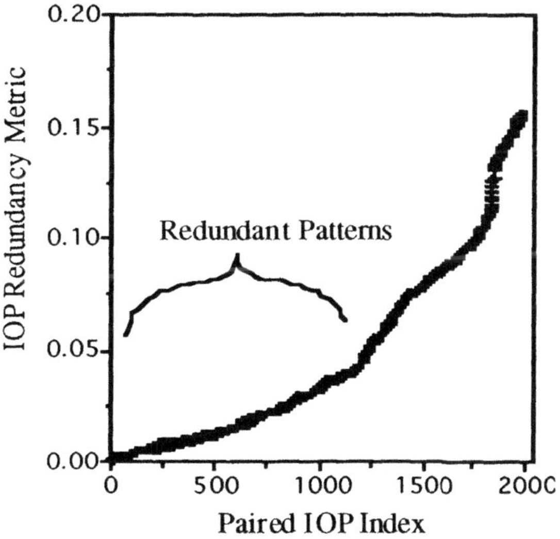

Figure 5.18 Sample Redundant Training Data Analysis

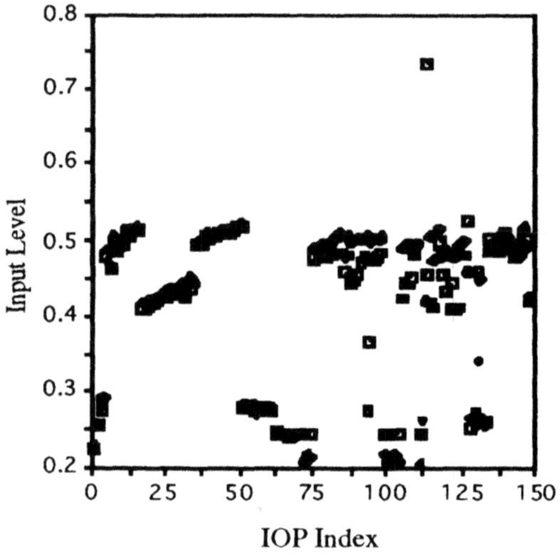

Figure 5.19 ANN Training Data Clusters

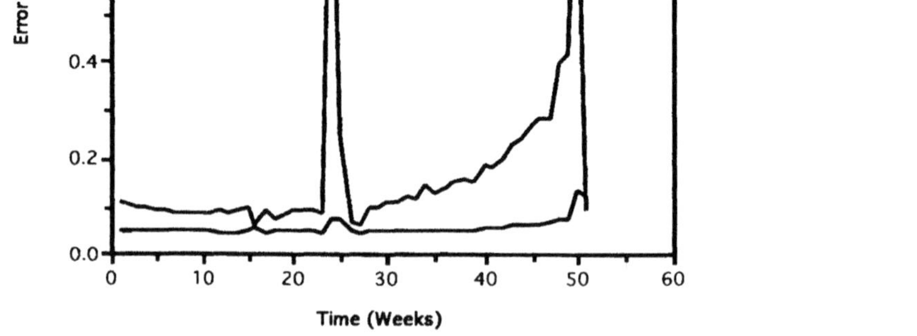

Figure 5.20 Output Errors for Cluster Boundary ANN

results using the minimum cluster boundaries are illustrated in Figure 5.19. As shown, the RMS error reached the level of 0.24, which is still too high for accurate state determination.

On closer inspection of the errors (target output - actual ANN output) associated with the correlated detectors 23 and 18, it can be seen (Figure 5.20) that the ANN had difficulty around week 51, which corresponds to an overall power drop. Additionally, this network misread a failed detector as a power drop, as indicated by the error for detector 18 at week 24. Therefore, we next consider the inclusion of power as an input to the neural network.

5.4.4 State Detection ANN with Clustered Data Boundaries and Power

From the results obtained previously in section 5.4.3, it is evident that the overall power level should be included as an input to the neural network. In this manner, the ANN will be able to compare the flux changes with the power changes to determine whether a detector has failed or that it has changed dramatically due to a similar change in power level.

The overall power level is now represented as input nodes for the current (time 1) and previous (time 0) weekly data samples. The rest of the network remains the same, and training is continued with the previous network's weights as a starting point for all paths except the new paths created to accommodate the power inputs (initial values for these path weights are randomly chosen).

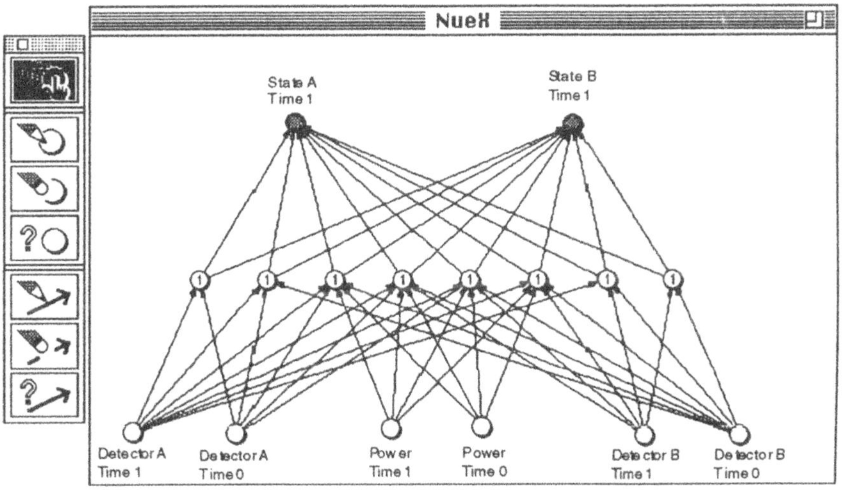

Figure 5.21 State Detection ANN with Power as Input

As illustrated in figure 5.22, the inclusion of power as an ANN input has allowed the network to converge to an RMS error of less than 0.05 after approximately 1400 training cycles (only the last 250 cycles are illustrated in figure 5.24). Given that the target ANN output levels are discrete, with 0.9 for failure states and 0.1 for normal states, an RMS error of 0.05 insures that the ANN has successfully classified all training patterns and therefore will generalize to similar patterns that it has not been trained with. The complete ANN evaluation results are presented in section 5.7.

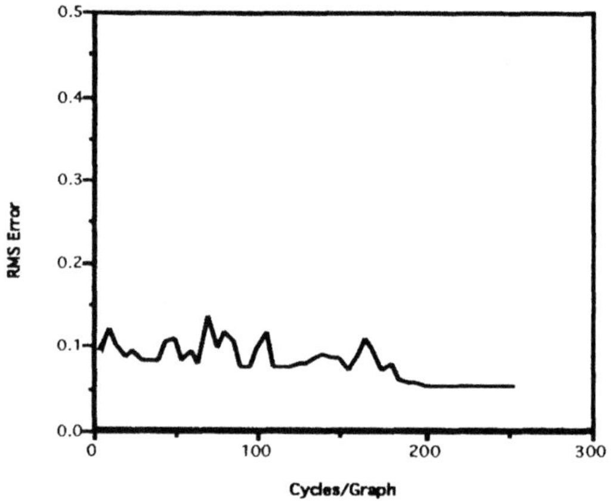

Figure 5.22 RMS Training Error using Power as ANN Input

5.5 Expert System Development

5.5.1 ANN Interpretation Knowledge Base

The ANN Interpretation Knowledge Base maps the outputs of the ANN into specific detector states. This is done by comparing the values for the two ANN output nodes to two thresholds: a high threshold used to detect detectors failures and a low threshold for normal detector conditions. The specific rules implemented for the ANN Interpretation KB are shown in Table 5.2.

Both detectors are declared Normal if their respective ANN outputs are less than the low threshold as asserted in rule 1. Rule 2 asserts a Both Fail condition when both outputs are greater than the high threshold. Rules 3 and 4 deal with one detector in the Fail mode and the other Normal. Rule 5 interprets both as Unknown when their respective ANN outputs fall between the high and low thresholds. Rules 6 and 7 are the case with one Unknown and one Normal. Finally, rules 8 and 9 interpret one Fail and one Unknown.

5.5.2 Formalized Majority Voter Algorithm

The detection of reactor core sensor hard failures is performed by the Majority Voter Knowledge Base. Hard failures, as opposed to soft detector failures, are characterized by large deviations from nominal values in detector flux measurements or by complete detector dropouts. The approach employed in the Majority Voter KB to detect these failure states is based on the Formalized Majority Voter algorithm presented in Lorczak, et al., 1989. In this section, we begin by describing the voter algorithm, and then present a description of its KB implementation.

Formalized Majority Voter Algorithm

The Formalized Majority Voter algorithm is a generalized voting technique applicable to arbitrary outputs using a metric space approach. In particular, it allows an inexact notion of equality among a set of outputs, or in this case sensor measurements, via the use of a pre-defined threshold. The main idea behind the algorithm is to choose a majority set from the overall output set in which all elements of the majority set agree within the pre-defined threshold with each other. We begin by defining a distance metric on a output space **X** and denote this metric d. Recall that d is real-valued function defined on the Cartesian **X** x **X** product space and with the following properties:

(i) $\mathrm{d}(x,y) \geq 0$

(ii) $\mathrm{d}(x,y) = 0$ implies $x = y$

(iii) $\mathrm{d}(x,y) = \mathrm{d}(y,x)$

(iv) $\mathrm{d}(x,z) \leq \mathrm{d}(x,y) + \mathrm{d}(y,z)$

where x, y, and z are in **X**.

The procedure for the Formalized Majority Voter can be stated as follows:

(1) select an element x from output set **A**

(2) let set $\mathbf{S} = x$

(3) select y from set $\mathbf{A} - \mathbf{S}$

(4) if $\mathrm{d}(y,z) \leq \varepsilon$ for all $z \in \mathbf{S}$ then $\mathbf{S} = \mathbf{S} \cup \{y\}$

In other words, the algorithm forms from the output set, in our case the set of n correlated sensor measurements, subsets in which all elements are within the threshold of each other. The set with the greatest number of elements is chosen as the majority set. For our application, any measurement not within the majority set is taken to be in failure mode.

Table 5.2 ANN Interpretation Knowledge Base

Rule No.	IF	THEN	Rule Firing* Priority
1	ANN output A< Low Threshold & ANN output B< Low Threshold	assert State A is Normal and State B is Normal	Medium
2	ANN output A> High Threshold & ANN output B> High Threshold	assert State A is Both Failed and State B is Both Failed	Medium
3	ANN output A< Low Threshold & ANN output B> High Threshold	assert State A is Normal and State B is Failed	Medium
4	ANN output A> Low Threshold & ANN output B< High Threshold	assert State A is Failed and State B is Normal	Medium
5	ANN output A> Low Threshold & ANN output A< High Threshold ANN output B>Low Threshold & ANN output B<High Threshold	assert State A is Unknown and State B is Unknown	Medium
6	ANN output A > Low Threshold & ANN output A< High Threshold ANN output B< Low Threshold	assert State A is Unknown and State B is Normal	Medium
7	ANN output A< Low Threshold & ANN output B> Low Threshold ANN output B< Low Threshold	assert State A is Normal and State B is Unknown	Medium
8	ANN output A> High Threshold & ANN output B> Low Threshold ANN output B< High Threshold	assert State A is Failed and State B is Unknown	Medium
9	ANN output A> Low Threshold & ANN output A< High Threshold ANN output B> High Threshold	assert State A is Unknown and State B is Failed	Medium

KB Implementation

Table 5.3 lists the rules implemented in the Majority Voter KB. The five rules are listed in order of rule firing precedence, i.e., which rules are interrogated first during the inferencing protocol. Rule 1 performs steps 1 and 2 by creating sets **S** and **Y** from the output set **A**, given that we have not exhausted all elements of **A** in the procedure. Rule 2 fires when both sets **S** and **Y** are not empty and chooses an element y from **Y**, then calculates the distance metric (dist) between this element and all elements in **S**. The distance metric used is the usual distance function on the real line. If the distances, given by the variable dist, are within the pre-specified threshold e then rule 3 asserts that element *y* is added to set **S**. Rule 4 checks that all elements of **Y** have been interrogated. If so, set **A** becomes **A** - **S**, a temporary set **S_temp** is created which is equal to **S**, and **S** becomes the null set. If **S_temp** exists and if the number of elements in **S_temp** is greater than the majority set, then **S_temp** becomes the majority. This logic is implemented in rule 5. The inferencing continues until the set **A** is empty.

Table 5.3 Hard Failure Knowledge Base

Rule Number	IF	THEN	Rule Firing * Priority
1	Set **A** is not the null set & Set **S_temp** is the null set & interrogated all elements of **Y**	assert Set **S** consists of the first element of **A**, Set **Y** = **A** - **S**	Very High
2	Set **S** is not the null set & Set **Y** is not the null set	assert choose *y* from **Y**, dist is the distance between *y* and all elements of **S**	High
3	Set **A** is not the null set & $\max(\text{dist}) <= \varepsilon$	assert add *y* to Set **S**	High
4	interrogated all elements of **Y**	assert Set **A** is **A** - **S**, **S_temp** is **S**, **S** is the null set	Low
5	**S_temp** is not the null set & number of elements in **S_temp** > **Majority_Set**	**Majority_Set** is **S_temp**, **S_temp** is the null set	Low

*ordering precedence for firing of rules

P8892-071

5.5.3 State Isolation Knowledge Base

The isolation of detector failure states based on ANN results is performed in the ANN Isolation Knowledge Base. For a selected detector, this knowledge base interrogates ANN results as determined by the ANN Interpretation KB for all the detector pairs having the currently selected detector as a member, and also looks at the previous state of that detector as determined by the Assessment KB to make a state isolation decision. The two possible results of this KB are: Fail, in which the detector is determined to be in a failure state (either a Soft Fail or Gradual Fail) and Normal, which relates to a normal operating state.

For the Maine Yankee detector positions, each detector is correlated with up to three other detectors. Therefore, up to three results from the ANN

Interpretation KB are available at each instance. For detectors with less than three correlations, the remaining ANN Interpretation KB outputs are set to Unknown. Each correlated pair presents four possible outputs: Unknown, Normal, Failed, and Both Failed (where both detectors in the correlated pair produced ANN outputs representing failed states). Adding the previous Assessment KB result of either Fail (whether a Gradual, Soft, or Hard) or Normal into the reasoning, produces 40 possible conditions on which to determine the current detector state.

Table 5.4 ANN Isolation Knowledge Base

Rule Number	IF	THEN	Rule Firing* Priority
1	Previous Assessment is Normal & Fail not in Detection & Detection contains at most one Both Fail & Detection is not Both Fail, Unknown, and Unknown	assert Isolation is Normal	Medium
2	Previous Assessment is Normal & Fail is in Detection	assert Isolation is Soft Fail	Medium
3	Previous Assessment is Normal & Fail not in Detection & Detection contains more than one Both Fail	assert Isolation is Gradual Fail	Medium
4	Previous Assessment is Normal & Detection is Both Fail, Unknown, and Unknown	assert Isolation is Gradual Fail	Medium
5	Previous Assessment is Fail & Detection contains Normal & Detection does not contain two Both Fail	assert Isolation is Normal	Medium
6	Previous Assessment is Fail & Detection does not contain Normal	assert Isolation is Soft Fail	Medium
7	Previous Assessment is Fail & Detection contains more than one Both Fail	assert Isolation is Soft Fail	Medium

Pgg92-076

*ordering precedence for firing of rules

The specific rules implemented in the ANN Isolation KB are listed in table 5.4 in order of rule firing priority. The reasoning for the 40 possible combinations has been encapsulated within seven rules. All the rules are at the same firing priority, i.e., no specification is made on which rules fire first. The first four rules deal with the situation where the previous assessment of the selected detector is Normal, whereas rules 5 through 7 deal with a previous assessment of Fail. Rule 1 asserts an ANN isolation of Normal with the previous assessment of Normal and none of the three correlated pairs reading Fail, at most one pair reading both sensors failed, and not the combination of two unknown and one both failed. An assertion of Soft Fail is made in rule 2 with a previous assessment of Normal and at least one Fail from the three pairs. Rules 3 and 4 are used to identify gradual failures, i.e. failures that are characterized by a gradual divergence in detector levels. Gradual failures are asserted in rules 3 and 4 with a previous assessment of Normal and not one Fail and not more than one both failed condition or for the condition of both failed and two unknowns. With a previous assessment of Fail, a result of containing at least one Norman and at most one Both Fail results in a Normal ANN isolation in rule 5. This is to represent detector spikes from the failed to the no fail condition. Finally, rules 6 and 7 assert a Soft Fail from a previous assessment of Fail. Rule 6 asserts a Soft Fail with no Normal and rule 7 asserts with more than one Both Fail.

5.5.4 Assessment Knowledge Base

The overall assessment of detector failure states is performed in the Assessment Knowledge Base. The Assessment KB uses results from the Majority Voter KB and the ANN Isolation KB to make the overall hybrid system assessment of detector operating status.

In rule 1, if both the Majority Voter and the ANN Isolation KBs are declaring the detector in a Normal state, then the Assessment KB asserts Normal. A Soft Fail is asserted when the Isolation KB has asserted Soft Fail and Voter asserts Normal as given in rule 2. Rule 3 fires when a Gradual Fail is detected by the Isolation KB with again a Normal result from the Majority Voter. Rules 4 and 5 assert an assessment of Normal when the Majority Voter result is Normal and the correlated pair is registering a Majority Voter Hard Fail and a Detection of Both Fail or an ANN isolation of Gradual Failure and correlated ANN Isolation of Soft Fail. Since the primary function of the formalized majority voter algorithm is to detect hard failure states, a Hard Fail is asserted in rule 6 when the Majority Voter KB returns a Hard Fail.

5.6 Hybrid System Development

5.6.1 Hybrid Neural Network/Expert System Environment Overview

NueX is a hybrid environment within which neural network and expert system technologies are combined to develop intelligent applications that outperform either method alone. Expert systems are useful in applications for which the expertise of a domain expert is available or the domain knowledge is structured and this expertise or knowledge can be described as a set of rules and facts. Neural networks are useful in applications for which there is no structured domain knowledge. Most real life problems fall somewhere in between these two realms and, therefore, a hybrid environment which integrates both technologies desirable. *NueX* is such a hybrid environment, the functionality of which has been enhanced through this Phase I effort.

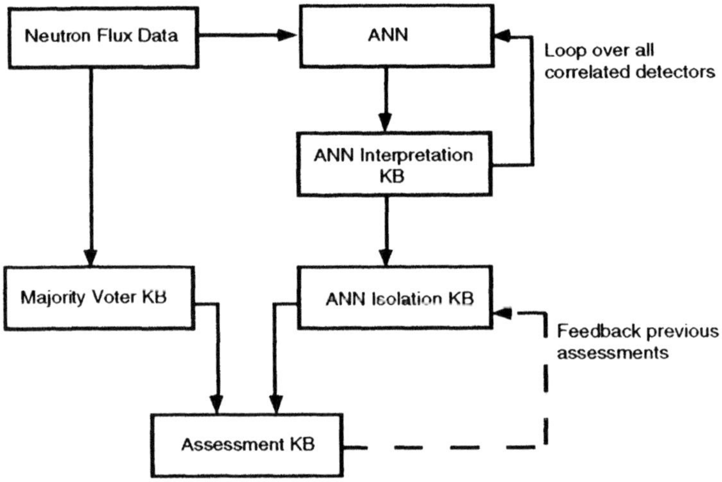

Figure 5.23 Hybrid Nuclear Monitoring System Architecture

NueX provides the user with the functionality to create hybrid applications employing both knowledge based Expert Systems and Artificial Neural Networks. This functionality is provided with the development of HyperCard stacks within *NueX*. These stacks allow the user to access ANNs and knowledge bases developed within *NueX* and incorporate all the graphical capabilities in HyperCard. The nuclear monitoring system prototype documented in this report is an example hybrid application developed within *NueX*.

5.6.2 Nuclear Monitoring System Architecture

The prototype nuclear monitoring system is a hybrid integration of the neural networks and expert system knowledge bases described in the previous sections. Within this hybrid architecture, neural networks are employed to detect and isolate flux detector failures which consist of subtle temporal and spatial changes. Knowledge based expert systems are used to determine more drastic detector failures, interpret the ANN results, and provide overall monitoring assessments.

The prototype hybrid system architecture, shown in Figure 5.23, illustrates how the developed ANNs and KBs work together to provide intelligent monitoring capabilities. For each weekly data set available, the State Detection ANN is presented flux measurements for each correlated detector pair, while the ANN Interpretation knowledge base continually interprets the ANN results. When the interpreted ANN results have been collected into a database, the ANN Isolation knowledge based evaluates the results, taking into account the fact that some detectors are correlated with more than one other detector. This ANN Isolation knowledge base also incorporates each detector's previous state assessment, in order to logically isolate failure states to the appropriate detectors. Concurrently, the Majority Voter knowledge base searches through the flux data to determine if Hard failure states have occurred, which the ANN has not been trained to account for. The Assessment knowledge base makes the final determination of each detectors operating state, illustrates the results within the graphical user interface, and compares that assessment with the detector's known state for system evaluation purposes.

5.6.3 Prototype Application Interface

Within the *NueX* hybrid environment, neural networks, knowledge bases, and data files are all opened and become automatically available to the hybrid application interface. The nuclear monitoring prototype system interface is implemented within *NueX* and provides adaptability for various nuclear plant core designs. Shown in Figure 5.24 is the system interface for the Maine Yankee core design. Detectors, labeled numerically within associated core positions, can be added or removed from the core through a simple menu option. Upon adding detectors, the interface automatically determines all the detectors which are symmetrically correlated, and lists the appropriate correlated detector pairs.

Illustrated in the figure, detector 10 (level 1) has been assessed as a Hard failure, while detectors 13 and 18 are assessed as Soft or Gradual failures. Each subsequent core level can be visually investigated for detector assessments. The *Show Results* button presents the complete data base which the hybrid system continually updates during execution.

5.7 Nuclear Monitoring System Evaluation

Here, we document the performance of the prototype Hybrid Nuclear Monitoring System. Our goal is to demonstrate the feasibility of such a hybrid system approach to nuclear plant monitoring. Specifically we look at the closed-loop performance of the prototype system in detecting failure states for the neutron flux measurements. Closed-loop refers to the system using its own previous results to make determinations on current detector status. The closed-loop performance of the following system elements are documented:

- Overall hybrid architecture performance as given by the Assessment KB
- Majority Voter KB performance in detecting hard failure states
- ANN performance as given by the Detection and Isolation KBs

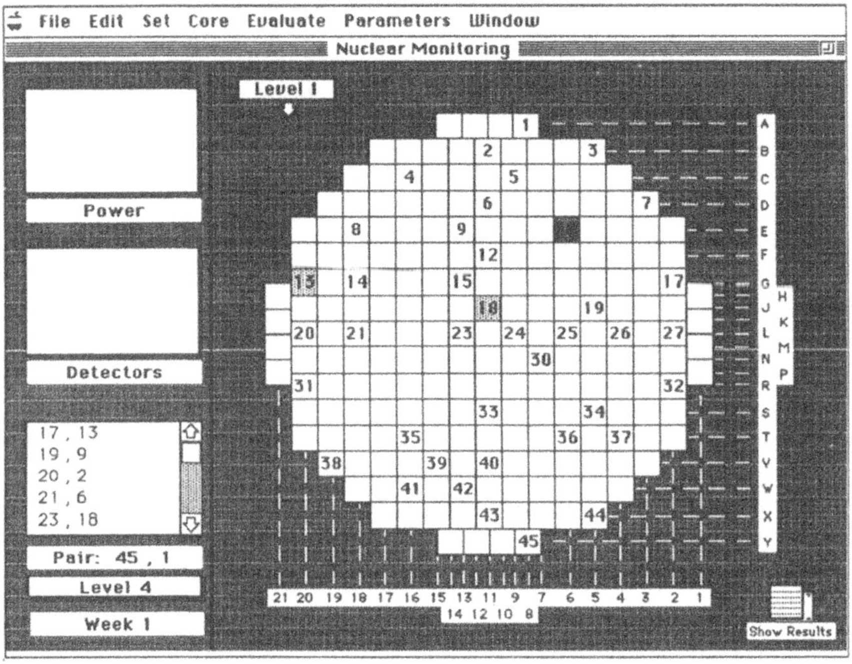

Figure 5.24 Nuclear Monitoring System Interface

Performance is judged based on three metrics: 1) probability of missed detection (P_{MD}) of failed states; 2) probability of false alarm (P_{FA}); and 3) average detection time for failed states (t_D). The specific definition of these metrics is detailed in Table 5.5. As shown, a missed detection occurs if a the hybrid system does not detect a Fail state over the entire time span for which a failure is present. A false alarm is generated whenever a Fail state is detected by the hybrid system, though it is known that the detector is in a Normal state. The average detection time (t_D) is the average time of detection over all the failure states classified correctly. It is measured in time intervals from failure onset to failure detection. Closed-loop hybrid system results are generated for 40 neutron detector measurements over a 51 week period. Table 5.6 shows the statistics generated from the performance evaluation.

Table 5.5 System Performance Metrics

METRIC	CALCULATION
Probability of Missed Detection (P_{MD})	$\frac{\text{No. of Fail States not Detected}}{\text{No. of Fail States}}$
Probability of False Alarm (P_{FA})	$\frac{\text{No. of Normal States Detected as Fail}}{\text{No. of Normal States}}$
Average Detection Time (t_D)	$\tau_D = \frac{1}{\text{No. of Fail States Detected Correctly}} \sum_i \tau_{D_i}$

Table 5.6 System Performance Evaluation

Performance Metric	Performance Level	No. of Samples
Prob. of Missed Detection (P_{MD})	1.89%	53
Prob. of False Alarm (P_{FA})	1.37%	6560
Average Detection Time (t_D)	1.35 (weeks)	52

Table 5.7 Individual Component Performance Evaluation

Failure State	P_{MD}	P_{FA}	t_D
Soft Fail	3.33% (1/30)	1.23% (8/6560)	0.14
Gradual Fail	0.00% (0/7)	0.11% (7/6560)	9.43
Hard Fail	0.00% (0/16)	0.03% (2/6560)	0.00

As shown, there are 53 failure states over the selected time span. The resulting missed detection rate is 1.89%, resulting from one failure not detected. We also see an exceptionally low rate of false alarms, 1.37%, which corresponds to 90 false alarms over the 6,560 samples available. For the 52 failures detected by the hybrid system, the average detection time was just over one time unit (one week), indicating the system processed failures almost immediately. These results clearly indicate that the Hybrid Nuclear Monitoring System is performing at a sufficiently high level to demonstrate feasibility.

In the previous paragraph, we investigated the overall hybrid system performance. We now document how the individual hybrid system components perform. Specifically, we address the performance of the Majority Voter KB in detecting hard failure states and the ANN performance in detecting soft and gradual failures states as given by the Interpretation and Isolation KBs.

Table 5.7 lists the results for the three different failure states: Soft Fail, Gradual Fail, and Hard Fail. Listed for P_{MD} and P_{FA} are the percentages, and in parentheses the number of occurrences of the states over the total number of samples available. The first two failure states relate to the performance of the Detection and Isolation KBs, whereas the latter state relates directly to the Majority Voter KB. The only missed detection was a soft failure. All hard and gradual failures were detected. False alarms were mostly due to soft failures. Average detection times, t_D, for soft and hard failures were very small, whereas for gradual failures they were higher, as one would expect. It is more difficult to quickly detect a slowly diverging detector than it is to detect temporal variations that are characteristic of soft failures and hard failures. These results thus point out that the hybrid nuclear monitoring system is performing very well in identifying off-normal operating states.

Table 5.8 demonstrates how individual component performance can be ascertained from system results. Shown are the week 21 results for several detectors. Each system output file contains seven columns. The first two columns identify the detector and the level. Columns three through six give the results from the Interpretation, Isolation, Majority Voter, and Assessment KBs, respectively. Column seven indicates whether the hybrid system detected the state correctly as given by an OK or Incorrect. For example, Detector 2, Level 3 shows the Interpretation KB outputting B (or Both Fail) for the three correlated pairs. The Isolation KB takes this information and asserts that a Soft Fail has occurred. However, the Majority Voter asserts that a hard failure has occurred. The Assessment KB then correctly asserts the Hard Fail state.

Figure 5.25 shows an example of system performance, giving the flux measurements for a pair of correlated detectors in which a gradual failure occurs. The correlated pair consists of the third level of detectors at core positions 14 and 39. Detector 14 (solid line) is failed from the onset.

The Hybrid Nuclear Monitoring System detects the failure at week 12 as pointed out in the figure. Therefore, the State Detection ANN has successfully identified the Gradual Fail state when there is no major temporal variation. The Majority Voter algorithm, and related filtering techniques, have considerable difficulty with very subtle problems of this nature. This example clearly shows the advantage of using neural networks for nuclear process monitoring. However, since the Majority Voter KB handles the relatively easy Hard Fail

Table 5.8 Sample of Hybrid System Results

Detector	Level	Detection	Isolation	Voter	Assessment	Status
Week21						
Detector1	Level1	N	(Normal)	(Normal)	(Normal)	OK
Detector2	Level1	N,N,N	(Normal)	(Normal)	(Normal)	OK
Detector2	Level3	B,B,B	SoftFail	HardFail	HardFail	OK
Detector2	Level4	N,N,N	(Normal)	(Normal)	(Normal)	OK
Detector5	Level1	N	(Normal)	(Normal)	(Normal)	OK
Detector5	Level2	B	SoftFail	(Normal)	SoftFail	OK
Detector7	Level4	B	SoftFail	HardFail	HardFail	OK
Detector9	Level3	B	GradFail	(Normal)	GradFail	Incorrect
Detector9	Level4	N	(Normal)	(Normal)	(Normal)	OK
Detector14	Level2	F	SoftFail	(Normal)	SoftFail	OK

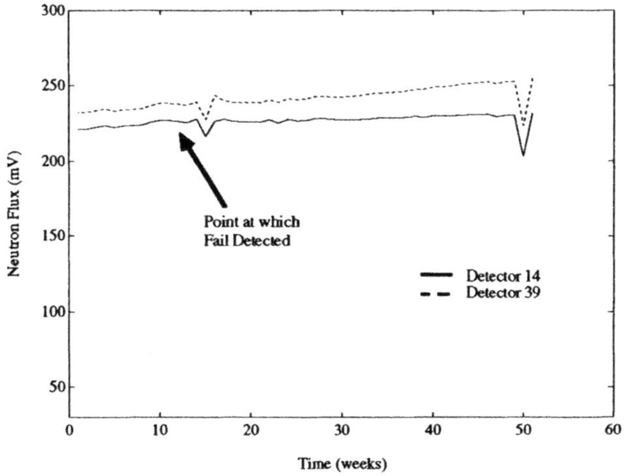

Figure 5.25 Time History for a Gradual Failure

states, the neural network is allowed to focus on the more difficult Soft and Gradual Fail states. Therefore, both neural networks and expert systems are working together to provide excellent overall system performance.

5.8 Conclusions

In our study, we developed a hybrid neural network/knowledge base-approach to intelligent sensor monitoring in nuclear power plant applications, developed a hybrid neural network/knowledge base plant monitoring architecture, implemented a prototype plant monitoring system using our neural network/expert system shell NueX on the Macintosh platform, and showed the feasibility of our approach by detecting failures in neuron detectors using in-core neuron flux samples provided by Yankee Atomic.

In our approach, neural networks classify the normal and off-normal operating states, while the knowledge bases interpret the ANN results, isolate off-normal states, and assess the situation with the expert knowledge encoded in a knowledge base. In particular, our investigation has shown the following:

- We have demonstrated that backpropagation trained feedforward neural networks can detect neutron detector failures. In addition, we have shown that time dependent paradigms are necessary for detecting failures for sensors without redundancy. The scope of our Phase I study covered *weekly* samples of in-core neutron flux samples.

- We have shown that single hidden layer feedforward networks have sufficient generalization capability in detecting neutron detector failures. We have also shown that power measurements are needed as ANN inputs in addition to moving window of the correlated detector measurements for the accurate detection and isolation of neutron detector failures.

- We have demonstrated that expert systems can complement the neural network based nuclear plant monitoring systems by detecting hard neutron detector failures and interpreting the ANN results.

- We have measured the performance of our hybrid neural network/expert system plant monitoring system using in-core neutron flux data provided by Yankee Atomic, and demonstrated its low false alarm and high probability of detection.

- We have developed a prototype plant monitoring system with a graphical operator interface using our hybrid neural network expert system environment *NueX* on the Macintosh.

Chapter 6

Chemical Tank Control System

To demonstrate the use of the network abstraction mechanisms in hybrid systems, James Hendler and Anne Wilson of the University of Maryland describe here an example using a system that controls the pH value of a controlled stirred chemical tank reactor. The system was created using the Conncert hybrid neural network/expert system development environment described in Chapter 2 and by Wilson and Hendler (1993).

6.1 Overview of the Chemical Tank Control Problem

The automated control of the pH value of a solution in a stirred chemical tank is a common problem. For example, this problem is found in off shore oil drilling. This drilling involves extracting usable oil from the drilled mixture and returning the remainder back to the ocean. For environmental reasons, this remainder must be in a non-toxic form. Automated methods are used to neutralize this solution before returning it to the ocean. Better control of such neutralization processes is a major factor in many chemical processes and environmental applications, and small differences in control can have major impacts in financial and environmental terms.

To test the use of a hybrid system in chemical process control, we are using a simulator that models a highly nonlinear chemical tank control system. In our model, inputs to the tank include an uncontrolled flow of an acidic solution and a flow of a base solution that comes through a controlled valve (see

Figure 6.1). The pH value of the acid may vary and additional disturbances may occur. The goal of the control system is to adjust the valve on the base solution to keep the pH of the stirred mixture within the tank at a constant value (called a set point). The particular tank, acid and base solutions we use are chosen to provide a highly non-linear control task, with a high rate of change near the point of maintaining a neutral (pH = 7) solution.

The traditional way to control such a tank is via the use of a "Proportional plus Integral plus Derivative," or PID, controller. Such controllers, which have been in common use for over forty years, are the most widely used method for control in chemical process plants. However, while the PID controller alone is sufficient to control the pH of such tanks, in the case of disturbances in nonlinear systems, the time in which it takes to return to a set point may be longer than optimal.

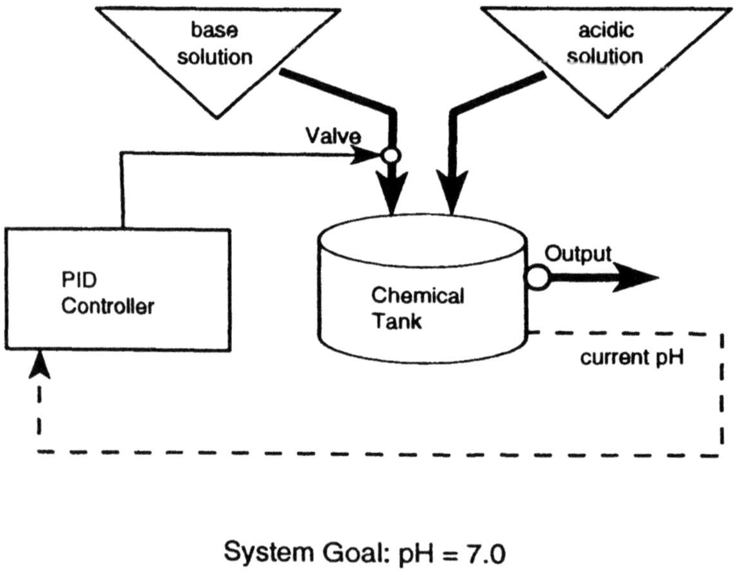

Figure 6.1 Schematic diagram of a chemical tank with control mechanisms

6.2 The PDP Hybrid Neural Network/Expert System

Improving on PID control is the "better mousetrap" of chemical process control. Particularly beneficial are methods that can use past data to improve the system, rather than requiring the development of correct equational models. One current method suggested for improving on PID control is the addition of neural networks into the control system. One example of this is the recent work of Cui, Hendler, Su, and McAvoy (1993), which improves the control of a stirred tank system by continuing to use a PID controller, but also adds a back propagation network that serves as a predictor of future tank pH values. The authors called this hybrid system the PDP ("Proportional plus Derivative plus Predictive" Controller.) The system is illustrated in Figure 6.2.

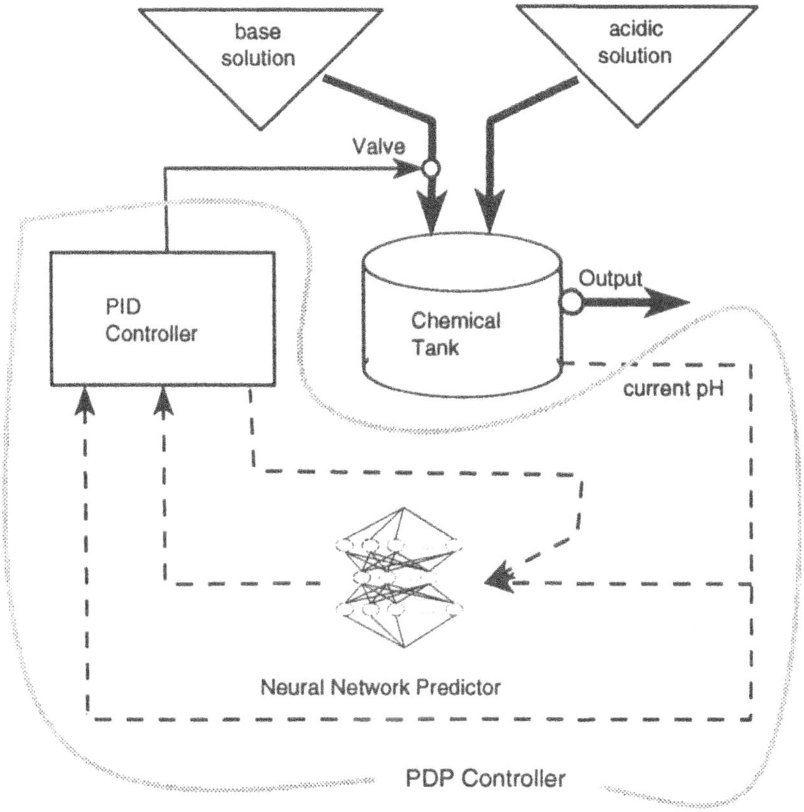

Figure 6.2. Chemical tank control hybrid system using neural network predictor

In the PDP system, the network takes as input a history of PID controller input values and tank pH values, and outputs expected pH values of the tank for later time steps. A disturbance is considered to occur when the tank's pH value is out of some bound and the network predicts that it will continue to go even further. At such a time, the network's prediction is included as an additional input to the valve controller, which performs a weighted average of the PID value and the predicted value to obtain better control.

If the predicted tank values were continually incorporated into the controlling equation, a possible outcome is "over control" of the system -- the system tries to compensate for predicted outcomes that are themselves a result of corrective action. For example, in the course of corrective action the neural network may predict a wide divergence from the set point in the opposite direction, causing the controller to take inappropriate corrective action in the wrong direction.

For this reason, a simple expert system is used to control when the predicted outcome is to be actually included in the controlling equation. The expert system examines the prediction and the current behavior. Under conditions having no disturbances, predicted values are not incorporated. If the plant appears to be moving away from set point, and the predictor forecasts a continued motion in that direction, then the prediction is averaged with the PID output and thus larger control actions (such as shutting the valve faster) will occur. More importantly, when the system's response and the prediction both are heading back towards the set point, the expert system will turn control back to the PID without prediction, letting the conservative nature of the PID algorithm dominate as the set point is reached. This is to avoid over control of the system and prevent the sort of oscillatory behavior that inclusion of the neural network prediction could cause.

The expert system uses a combination of temporal pattern matching (see Hendler and Dickens, 1991) and rules to determine when to intervene in control. To determine if the predicted value should be incorporated, the system determines whether a set of conditions holds. Somewhat simplified, the system must determine that

- the current system output is past some threshold away form the set point (this controls for the noisy inputs found in many plants)

- based on current and several past system outputs, the system can determine that the plant output is diverging form the set point

- based on the neural network prediction, the system output will continue diverging from the set point.

A second set of conditions is used to determine when the predicted value should cease to be included in the control (i.e., when the control can be returned to the PID controller without interference):

- the predictor must currently be on
- based on current and past outputs, the chemical system is returning toward the set point
- based on neural network predictions, the system's output will continue to move towards the set point.

Empirical testing of their system has shown that the inclusion of the predicted pH value causes the system to return to within the desired pH range faster than when using the PID controller alone. One example of the results of such control can be seen in Figures 6.3 and 6.4, which show the tank pH value during a disturbance both without and with the neural network predictor respectively. As can be seen, the set point is reachieved more quickly, and with less total disturbance, when the hybrid controller is used than when the PID is used alone.

6.3 Drawbacks in the Construction of the Hybrid System

The manner in which the back propagation network was incorporated into the hybrid version of the system was not atypical for inclusion of back propagation networks. First the network was constructed and trained in a process external to the system. The code developer oversaw the training process, adjusting parameters and watching for error minimization and convergence. Once trained, the network was tested using yet another separate process. Finally, the resulting weight matrix was hard wired into the control code by hand.

There are several drawbacks to this integration approach. First, the training and testing processes are separate and distinct from the operation of the system, even though correct system performance depends on the successful outcome of these processes. Nowhere in the code is there anything to reflect the state of the network or the quality of its output, such as whether it is trained or untrained. Instead, responsibility for ensuring that the network is in a usable state falls outside the scope of the system and on a person, the code developer.

The separate training processes and the hard coded weight values imply that the environment in which the system runs is static. Yet, it is very possible for the environment to change. For example, a shift in the acidic solution (changing the type of acid, as opposed to just the concentration) may occur,

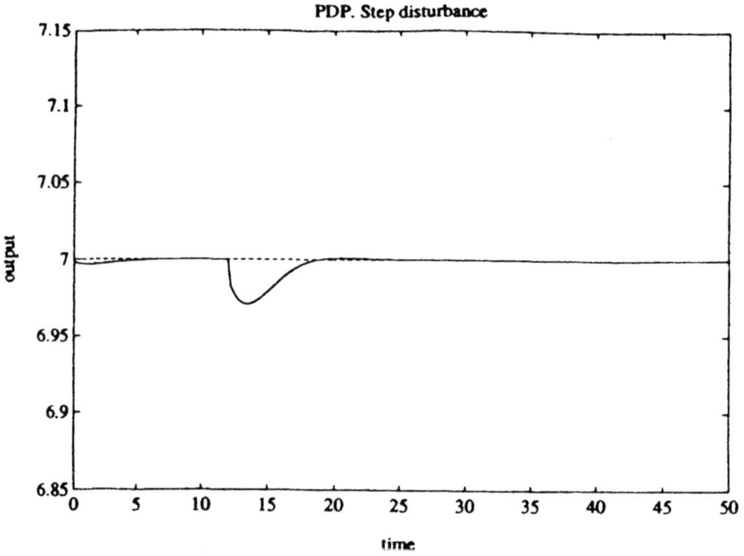

Figure 6.3. Tank pH value during a disturbance -- without the neural network predictor.

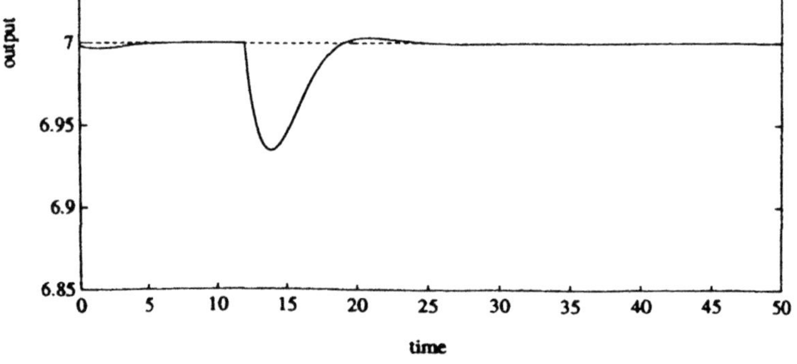

Figure 6.4. Tank pH value during a disturbance -- with the neural network predictor.

changing the composition of the mixture in the tank. In such cases, the previously trained network would no longer reflect the new environment.

The responsibility for detecting and responding to such changes again falls on a human monitor. Most likely, the system would be halted and off line training and testing processes would be repeated. During that interval the PID controller system without the network predictor may be run, although someone or something must initiate it. After retraining, the hybrid system code would be modified, with the code developer replacing the old weight values with the new ones by hand. Finally, the hybrid system would be restarted. In this scenario, all stopping and starting of processes must be done by some person, outside the scope of the system.

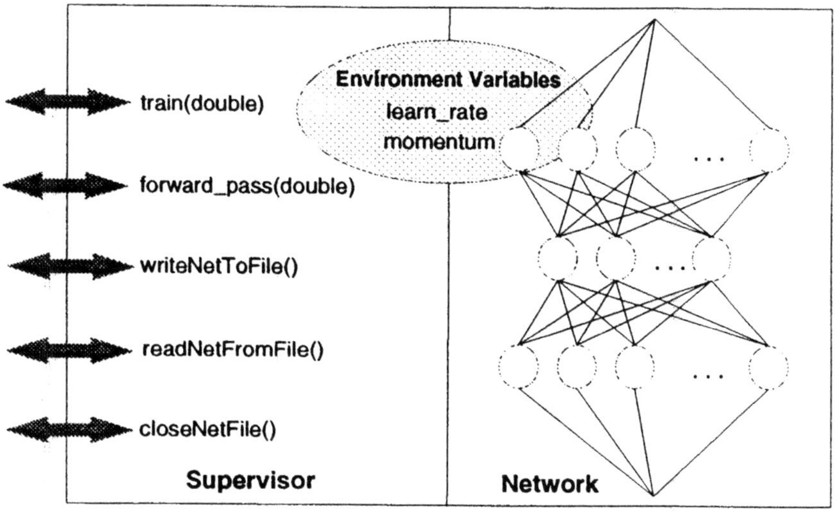

Figure 6.5. Conncert back propagation network object

Should the system designer decide to substitute some other prediction method in place of the back propagation network, it would be necessary to remove the weight matrix from the code and include references to the new method in the proper locations. The person making the modifications could only hope that the initial code developer followed good software design practices in writing the code, grouping related data structures and functions together so that this task would not be difficult.

6.4 Improving the Chemical Tank Control System with Conncert

The chemical tank control system code has been improved by replacing the back propagation weight matrix by a back propagation network object created in the Conncert development environment, which is described in Chapter 2 and in Wilson and Hendler (1993). The diagram of the Conncert component used to develop this application and the (simplified) corresponding object definition are shown in Figures 6.5 and 6.6. Note that the **public** part of the object definition defines the object's interface, as presented in the diagram. The Conncert implementation of the hybrid system is shown in Figure 6.7.

```
class BPNetObj : public VariedDimen {
protected:
        BP_Net myBPNet;
        virtual void doTrainMsg (message*);
public:
        BPNetObj (char* name, int nInputs, int nHidden, int nOutputs);
        int train(Int windowSize);
        int forwardPass();
        int writeNetToFile();
        int readNetFromFile();
        int closeNetFile();
};
```

Figure 6.6. PBNetObject definition

The back propagation network object needed for this particular application was instantiated from a class designed specifically for this problem, called a **pdpBPNetObject** (from the name PDP Controller). Figure 6.8 illustrates the definition of the **pdpBPNetObject**. The class **pdpBPNetObject** was derived from the more generic class **BPNetObject** shown in Figures 6.6 and 6.8. Thus it has the same basic functionality as the **BPNetObject** class, but with some modifications specific to this application. In particular, the training function has been rewritten specifically for this problem, and a new variable has been added to tell the supervisor how much of a history to use as input to the network. Since no additional functionality has been added, the interface is the same as for the **BPNetObject**.

The network is included in the system by the following declaration:

pdpBPNetObject myBPNet (BPNET, nInputs, nHidden, nOutputs, histBufferSize);

The first argument is a string giving the object a name. The next arguments give the number of input and hidden and output units. The number of output units reflects the number of future time steps the net will predict. The last argument indicates how big the window of past pH and PIDvalues will be. (The number of inputs is always two times the histBufferSize.)

When this network object declaration is processed by the compiler, the data structures for the network are created and the supervisor is spawned. At that time, the supervisor looks for a particular file that represents a back propagation network and associated state information. If that file exists, then the stored weight values are used to initialize the network, and the supervisor notes whether the net was labeled as trained or untrained. (A stored network may be only partially trained. This is particularly useful in cases such as this when training may take a very long time.) Otherwise, random weight values are chosen, and the supervisor considers the net to be untrained.

Generally speaking, in the case of an untrained network, the course of action taken by the supervisor may vary depending on the application. In this case, the supervisor will automatically put the network through a training phase. In this particular system, it is possible for the PID controller to function without the network albeit at less than optimal performance, so the system can continue to run while the net trains itself. If a request for output is made to the network while it is considered untrained, output is returned but the supervisor returns a confidence value of 0 and indicates via the error code that the network is not trained.

An alternative to having the encompassing system continuously query the net and repeatedly get back no confidence values, is to have the encompassing system maintain a net state variable, then have the super interrupt the system to change the value of that state variable. The difference is between continual polling and receiving an interrupt. Either one is possible and both may be offered.

Once the network is considered trained, the supervisor serves to monitor the network's performance. It does this by comparing the output of the net (the predicted future pH value) with the actual pH value at the next time step. If the difference is too large, the supervisor reduces the confidence value of the network's output. If the confidence value falls below some threshold value, the supervisor will put the net back into a training phase. In this application, data to train the network is always available since the input to the network is a window of past pH and PID values.

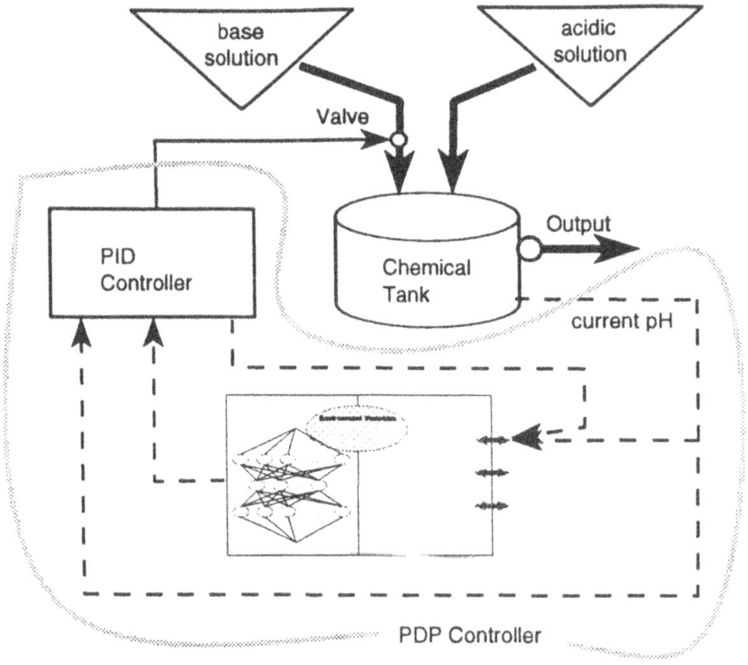

Figure 6.7. Chemical tank control hybrid using Conncert network object

```
class pdpBPNetObj: public BPNetObj {
      virtual void doTrainMsg (message*, double confidence);
      int histWindowSize;
public:
      pdpBPNetObj (char* name, int nInputs, int nHidden, int nOutputs, int
hWinSize);
};
```

Figure 6.8. The pdpBPNetObject definition

Sometimes the encompassing system may want to initiate retraining. A request for the network to train itself looks like this:

int result = myBPNet.train(2400, confidence);

In this particular implementation, the terminating condition of the training routine is defined as having 90% of the output vectors of a given window size all within an epsilon value of their corresponding target vectors. In this line of code, the parameter 2400 tells the training routine to use a window size of 2400 input/target pairs for the purpose of determining correctness.

Although we have reduced the dependency on human monitoring, these abstraction mechanisms have not completely mitigated that need, mainly due to network training. Although it is expected that the supervisor can train the network most of the time, it may not be able to do so always. We have included code that makes some assessment of the success of the training. For example, should the training take much longer than expected, the supervisor will detect this and report that information.

Thus, it is still necessary to have a person check the output of the supervisor from time to time to ensure that any training was successful. The difference is that the person need only check for messages from the supervisor, and doesn't need to actually trace the tank pH values to assess performance. In this sense, the supervisor has raised the level of monitoring from forcing the human observer to interpret a sequence of numeric values to giving the observer a message meaningful at the level of human discourse.

6.5 Future Enhancements

The chemical system explored by Cui, Hendler, Su, and McAvoy (1993) only includes one control loop; that is, the PDP controller is only used to control a single valve. In a more complex plant, much more complex regimes must be used. Such a plant motivated Tennessee Eastman Corp. to develop a simulator program that was released to the chemical process control community. The Eastman simulator requires controlling a plant with 41 measured values and 12 individually controllable valves. A proposed solution to steady state control of that plant (Ye, McAvoy, Kosanovich, and Piovoso, 1993), for example, requires four separate control loops, with two of them nested within others.

To apply the hybrid controller to more complex plants will require extensions in two directions: more complex predictions using the neural networks and an expert reasoner that not only decides when to add predicted values, but also how to balance different predictions against each other. To achieve the first of these goals, we plan to use a hierarchical design, requiring the use of Conncert's advanced features for handling modular neural networks (Wilson and Hendler, 1993).

Chapter 7

Image Interpretation Via Fusion of Heterogeneous Sources Using a Hybrid Expert-Neural Network System

This chapter was written by Naresh Gupta, Srinivasan Raghavan, Gretchen Bailey, and Laveen Kanal at the LNK Corporation, Inc. They describe the system InFuse, which addresses the multi-source integration problem.

The potential that underlies systems based on a hybrid of pattern recognition technologies is beginning to unfold rapidly. InFuse exploits the notion of combining expert systems and neural networks to capture the complementary advantages of both technologies. The application involved is to extract natural terrain features from imagery of multiple sensors in conjunction with terrain knowledge in the form of a geographical database. InFuse makes use of feedforward neural networks to interpret Gabor textural descriptors obtained from the imagery and an expert system to verify decisions from the networks with a geographical database. The power of our approach lies in its ability to combine symbolic data of the spatial database and domain specific knowledge about the sensor behavior with example-induced learning from neural networks to achieve a reliable classification of natural terrain features. This paper also discusses two different methods of sensor fusion referred to as data-level fusion and decision-level fusion. Results of applying both types of fusion are presented.

7.1 Introduction

Feature extraction from multi-source imagery and collateral sources, also known as terrain image understanding, can be of key importance in many military and civilian activities, including mission analysis, navigation, automatic target detection and recognition, spatial database updating, evaluation of environmental effects, and prediction of agricultural yield. Even though the desired features include both natural as well as man-made features, discussion in this paper is limited to locating only natural features. In order to detect these features in all weather/day conditions and improve reliability, more than one sensor can be used. Collateral data such as geographic or relational databases can be very useful. The amount of human effort required to make a decision using multiple sources can be enormous and time consuming. We find that a hybrid combination of neural networks and expert systems is very useful to automate the process of integrating heterogeneous data sources.

Terrain interpretation has experienced a substantial growth in the recent past. One of the popular approaches to terrain image understanding is to employ knowledge-based region labelling after application of a sequence of traditional image processing routines. The major advantage of knowledge-based interpretation of terrain imagery is the ability to make use of a vast body of contextual information which results in superior performance of the image understanding algorithms. However, this advantage comes with a large price, namely that introducing more scene labels will place a heavy burden on the knowledge base. Furthermore, it is difficult to conclude when the knowledge base is sufficient to guarantee correct interpretation of an arbitrary scene. Usually an expert system is used to drive the labelling process. Incorporating an exhaustive set of rules in the knowledge base is not a trivial task.

Neural network based fusion of multiple sources is recently becoming more popular. Rajapakse and Acharya (1990) proposed a five-layered neural network architecture for object recognition. It is not clear how this architecture will learn textural features since the principles of geometric pattern learning it uses are similar to those of Neocognitron. Comparison of neural networks and maximum likelihood classification has been obtained by Chettri *et al.* (1992) for interpreting remotely sensed data. Their work does not, however, address the issue of incorporating external sources of knowledge. In terrain classification *a priori* known sensor behavior is very useful to allow a variable gain control mechanism in weighing the input from different sensors. For example, in SAR imagery water regions stand out well consistently during all seasons due to the poor radar reflection, yet forests in SAR may not exhibit a consistent behavior due to absence of foliage in winter. Such important information can not be easily captured in neural networks.

Hybrid systems hold the key to the problem of integrating multiple heterogeneous data sources. Better performance results when hybrid systems are used to solve several complex problems (Kanal and Raghavan, 1992). This is because different computational tools are often found to possess complementary advantages. The notion of designing hybrid systems is not new. Twenty years ago in Kanal (1972) the following statement appeared:

> "It is now recognized that the key to pattern recognition problems does not lie wholly in learning machines, statistical approaches, spatial filtering, heuristic programming, formal linguistic approaches, or in any other particular solution which has been vigorously advocated by one or another group during the last one and a half decades as the solution to the pattern recognition problem. No single model exists for all pattern recognition problems and no single technique is applicable to all problems. Rather, what we have in pattern recognition is a bag of tools and a bag of problems."

Just as a skilled craftsman will not pursue his craft using just one tool, so must the pattern recognition engineer know how to use a combination of tools to attack a complex problem. In Kanal and Chandrasekaran (1972) the value of taking a hybrid linguistic-statistical approach over a strictly statistical or strictly structural approach was developed and illustrated with an example. Combining heuristic search with statistical pattern recognition and decision trees, and heuristic search with structural pattern recognition are presented in Kanal (1979) and Kanal and Dattatreya (1986). A variety of combined approaches to the design of decision trees is surveyed in Dattratreya and Kanal (1985). Also, combining structural and probabilistic knowledge with hypergraph search to develop alternate scenarios for decision making in uncertain domains is presented in Bhatnagar and Kanal (1992). While all this work stresses the important virtues of hybrid systems, it remains to be seen how such a system can be effectively used to addressed the problem of multi-source integration.

A close look at this problem suggests that it involves processing two entirely different types of data namely, the symbolic data about the sensor performance and local terrain knowledge from a spatial database, and the enormously complex sensor data. While symbolic information can be easily handled by expert systems, the complex sensor data poses a number of difficulties to the expert system-based interpretation of the aerial imagery. On the other hand, what appear to be complex patterns of the imagery can be quite adeptly handled by neural networks. We show that a robust and reliable feature extraction system results when a hybrid combination of neural networks and expert systems is used to fuse multi-source imagery and spatial databases. In the current implementation of the system, Synthetic Aperture Radar (SAR) and optical imagery were used to extract forests, water bodies, and fields by fusing SAR and optical imagery with the aid of an expert system and a geographic database. In fusing the imagery, we observe that two types of fusion are possible. One of them, known as decision-level fusion, relates to the concept of individually classifying the imagery first followed by fusing the class decisions at the next level. The other type, known as data-level fusion, relates to the notion of treating the textural features from all the images as a single input vector towards the goal of classification. Intensive treatment of the two types of fusion or discussion of theoretical aspects of the sensor fusion problem must be deferred to a later date following further investigayion. Here, in employing a hybrid framework for sensor fusion, we present our observations in experimenting with the system and some empirical conclusions derived from testing.

The remainder of this paper is organized as follows. We lay the foundation for the InFuse architecture in Section 2. One of the critical factors in deciding the accuracy of terrain classification is the faithfulness of the texture descriptor which characterizes the terrain. We devote a brief discussion on mixed spatial-frequency techniques to represent textures in Section 3. The two types of fusion and the expert system-based verification process is presented in Section 4. Implementation and results of testing are discussed. We conclude the paper with a few remarks on data fusion and hybrid systems in Section 5.

7.2 InFuse Architecture

Feature extraction requires extracting local components and then inferring or constructing the feature from its parts. Different sensors provide different types of information about features. By combining information from several sensors, the classification process can adapt to the strengths of each sensor type. Past research effort to automate the feature extraction process has been primarily on interpreting images from individual sensors. Much of this effort, as mentioned earlier, focused on knowledge-based interpretation of imagery using expert systems. A serious disadvantage of applying expert systems alone as the primary tool for feature extraction is that production rules are by nature constrained, often requiring a large number of rules to describe all variations. Thus, constructing rules to characterize every possible instance of a natural or man-made feature can be enormously difficult, especially when taking into consideration the presence of noise and variations in sensor behavior. On the other hand, neural networks offer the flexibility of having the network adaptively learn the regularities and irregularities of the input patterns, instead of requiring this information to be explicitly encoded as in expert systems.

Integrating these two technologies takes advantage of the pattern recognition capabilities of neural networks as well as the symbolic reasoning capabilities of expert systems in that their strengths complement each other. These technologies could be combined in an order that is most appropriate to the format of the input and/or output data. For example, sensor information is more readily accepted by the input nodes of a neural network, whereas characteristics about the sensors themselves that may affect the final classification lend themselves to object-oriented and production rule representations. In the case where sensor information must be used for classification, a trained neural network would accept the input sensor data and would pass its classification(s) to an expert system which would provide corrections and/or confirmations to the classification(s). An additional benefit to incorporating expert system technology in a neural network-based system for feature extraction and classification, is that the object oriented representation lends itself to the adaptation of data that resides in relational and/or Geographic Information System (GIS) databases. This expands the number and variety of information sources, which increases the depth and breadth of the domain.

The architecture of InFuse, as shown in Figure 7.1, is a three layered system in which an expert system is housed at the third level and a population of neural networks is configured to work at the second level, while a library of basic image processing and data representation schemes are made available at the first level. The three levels are loosely coupled and are allowed to interact with each other.

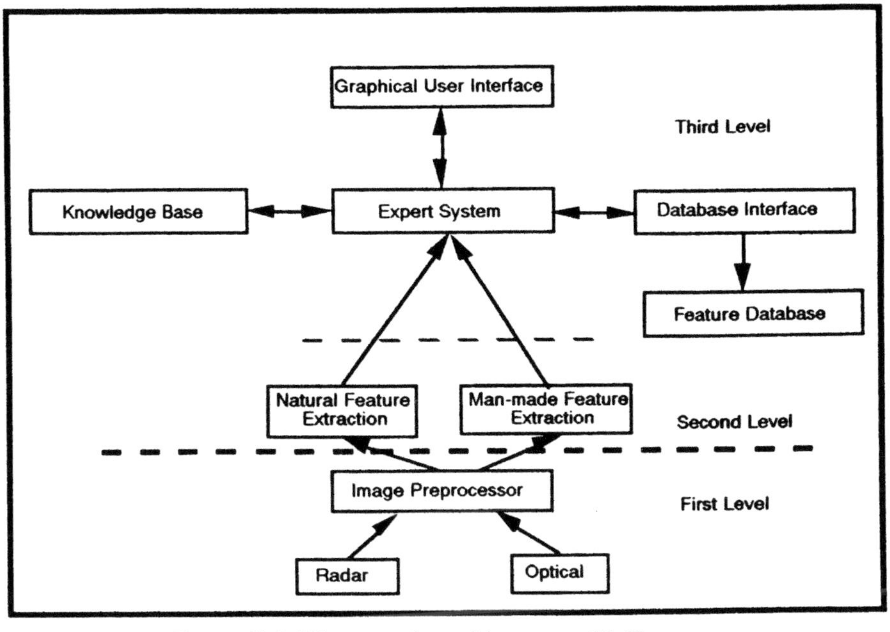

Figure 7.1 The generic architecture of InFuse

Beginning with the first level, the image preprocessor is equipped with a library of filtering schemes, image registration software, and image clean-up routines. The image preprocessor also contains routines to achieve discontinuity-preserved smoothing and edge detection (Blake and Zisserman, 1987) which form the input to man-made feature extraction. Edge features are of little use to natural feature extraction because textural variations of the natural features cause spurious noise. Instead, spatial frequency descriptions of textures are the most desirable texture descriptors. In this paper, we discuss a class of mixed spatio-frequency domain descriptors known as Gabor filters. Later, in the next section, we point out in detail why the choice of Gabor filters makes sense. For now, it suffices to say that an adaptive neural network to extract a Gabor representation of the image data is part of the first level image preprocessor routines.

At the second level, the neural network-based region feature extraction module has one or two stages of neural network processing depending on the kind of sensor fusion employed. In the case of data-level fusion, a collection of feed forward networks that use the Gabor filter responses from multiple sensors is arranged to "view" the entire image as a grid of cells to classify the image texture as forest, field, or water. In the case of decision level fusion, a second stage of processing is introduced to employ another set of feedforward networks for refining the decisions of networks from individual sensors to arrive at a collective decision on the image texture.

At the third level, the expert system is responsible for taking the output of the neural networks and refine the decisions made by the networks. In addition to analyzing the output of neural networks, the expert system can interface with a geographic database. This interface is designed to operate in one of the two modes: (1) using the database to improve feature extraction and recognition, and (2) refining the database using the output of the feature extraction system. The basic approach is to use the expert system to try to verify region boundaries as determined by the neural networks, with region boundaries as determined by the geographic database. The database is intended only to serve as a guide to the system, since such databases are often outdated and the contents should be tempered by more contemporary evidence. The expert system could be easily extended to provide feedback to the neural networks which would enable the expert system to task the neural networks to do additional processing for feature extraction. Thus inconsistencies in high level feature extraction could be used to re-focus lower level feature extraction with appropriate modifications to the neural networks to incorporate the additional contextual information available.

7.3 Data Representation and Feature Characterization

Feature extraction from multiple sensors involves the manipulation of potentially large amounts of data. In general, for many applications of object recognition, it is sufficient to consider the image information in the spatial domain. For instance, zero-crossings of the Laplacian convolved with a Gaussian of a range of standard deviations can represent the object as a set of edges (Marr and Hildreth, 1980). It may be noted that a complete signal can be recovered from its zero-crossings if the signal is bandlimited. However, this kind of representation is of little use for describing textures since the intensity changes of an irregular texture do not reflect the nature of the texture at all. In order for the neural networks to be able to learn a finite number of patterns, the input representation needs to be insensitive to two kinds of variations:

• The first kind of variation arises due to sensing conditions. For instance, a forest region in an optical image on a cloudy day will appear different on a bright sunny day if we consider using the grey levels of the pixels as the direct representation. Similarly, seasons play a considerable role in deciding the irradiance pattern of a feature in the image.

• The second kind of variation comes from the actual variations of appearance intrinsic to a feature. For example, roads may appear to be dark in color in one part of an image and light in another if cement is used instead of asphalt. It is clear that raw sensor data is sensitive to the sensing conditions in addition to being cumbersome to deal with.

An efficient data representation scheme, therefore, needs to be tolerant and somewhat insensitive to variations in imaging conditions as well as feature appearances, while preserving the locations where the image attributes are available. In order to extract textural information, one may consider Fourier domain representations for measuring the spatial frequency of the intensity patterns (in this case, the textural elements). However, a completely frequency domain characterization of textural information has no spatial location information of where the features appear.

One of the key aspects of our work is the use of mixed spatial and frequency domain representations such as the Gabor transforms to characterize local texture in an image. The use of Gabor transforms offers considerable promise in providing highly concise representations of textural features by emphasizing frequency information as locally as appropriate.

Over the past several decades, there have been numerous attempts to find suitable characterizations of texture. From the early days of texture analysis, frequency domain methods have played an important role in texture characterization. Due to their inability to characterize spatial location, in recent years there has been considerable interest in mixed time-frequency methods which provide a spatially localized signal characterization in the frequency domain. This is particularly useful for image analysis applications in which there are different textured regions in a single image. The Gabor filter representation (Daugman, 1988) is a good example of mixed time-frequency domain characterization. The Gabor transform has started to gain prominence because of its potential relevance to the receptive fields in biological vision. There is evidence that the family of 2-D Gabor filters captures important properties of the human visual receptive field including spatial localization, orientation, selectivity, spatial frequency selectivity, and quadrature-phase relationship.

The 2-D Gabor filters are a parameterized family of exponentially damped sinusoids. The sinusoids measure spatially oriented textures of different frequencies and the exponential damping provides localization. These filters are characterized by:

$$G(x,y) = \exp(-\pi[(x-x_0)^2a^2+(y-y_0)^2b^2]) \cdot \exp(-2\pi i[u_0(x-x_0)+v_0(y-y_0)]) \quad (1)$$

and the image can be represented using the Gabor filters as:

$$I(x,y) = \sum_i A_i G_i(x,y) \quad (2)$$

where $\mathbf{A_i}$ are the Gabor coefficients; $(\mathbf{x_0}, \mathbf{y_0})$ is the spatial location in the image around which the frequency information is obtained; and $(\mathbf{u_0}, \mathbf{v_0})$ are the filter orientation parameters. In an actual implementation, the image is divided into a grid of cells whose center is $(\mathbf{x_k}, \mathbf{y_k})$. The grid resolution and the filter frequencies can also be chosen arbitrarily. Figure 2 shows a reconstructed terrain image from three different choices of filter frequencies while the grid resolution is constant.

Some interesting properties of Gabor filters are worth mentioning. First, the Gabor representation has been shown to achieve maximal response in the conjoint spatial and Fourier domains. The direct implication of this property is the richness of information content in the Gabor coefficients for a given spatial grid resolution. Second, they can be viewed as forming the continuum spanning the two ends of Kronecker delta in spatial domain to Fourier representation in frequency domain. Third, the Gabor functions $\mathbf{G_i}$ are non-orthogonal and hence the Gabor coefficients need to be obtained through a gradient-descent procedure which minimizes the mean-square error in image representation (Daugman, 1988). Finally, the Gabor functions do not provide compact support and therefore overlapping is needed. Overlapped support places additional burden on the computational needs. For this reason, alternative ways to represent textures using multi-resolution based schemes such as wavelets can also be used. We are currently exploring these methods.

The separation of natural features from man-made features in our system bears an important relationship to the manner in which the Gabor coefficients are obtained. The Gabor coefficients are computed over a lattice of cells that occupy a region of pixels. The cells may or may not be lying on one single feature. In general, natural features occupy several cells making it advantageous to treat the textural behavior of an entire cell as the characterization of the dominant feature occupying that cell. Forests are characterized by a macro-level texture while fields exhibit a more finely-grained textural behavior in SAR images. In contrast, water tends to emit a constant intensity in SAR as well as in optical imagery.

Texture discrimination can be achieved by some form of normalized Gabor coefficients alone and thus these coefficients form the input to the neural networks trained to classify region features in our system. One of the major difficulties with using Gabor coefficients directly is that, in general, they have a strong d.c. component or high magnitudes for the zero frequency component filter. The strong zero frequency component dominates the direction of the input

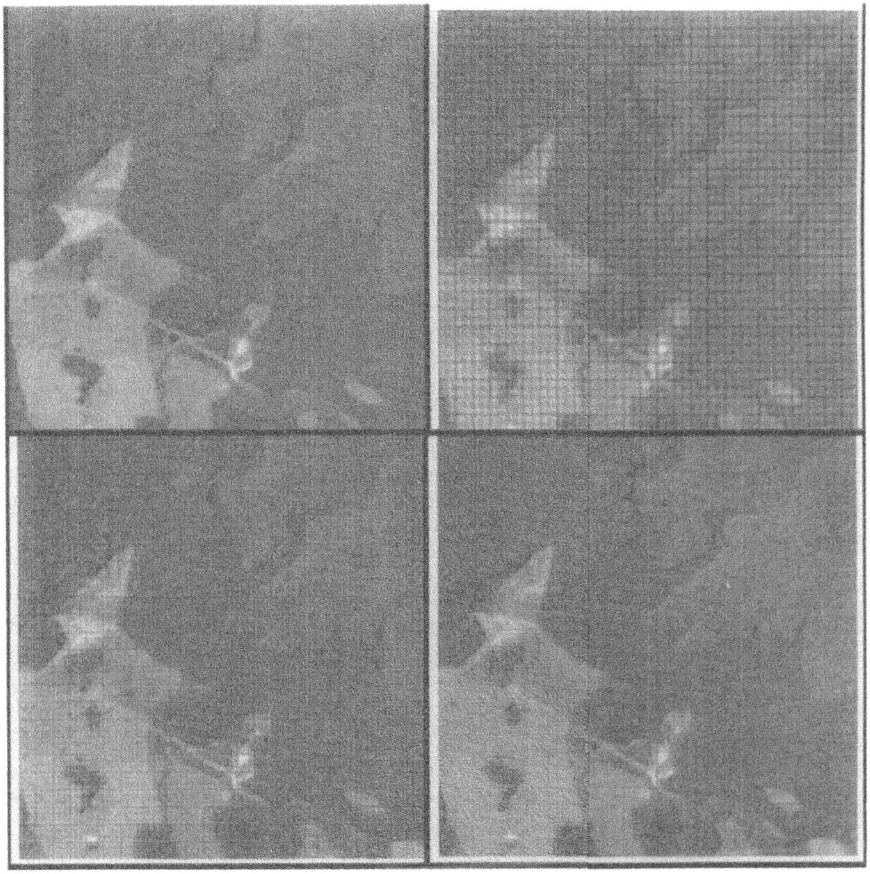

Figure 7.2 Gabor reconstructions: Original image sample is shown on the top left. Reconstructed images using Gabor filters are shown in the top right and bottom portions. The number of filters used are 9, 25, and 49 for the top right, bottom left and bottom right respectively.

vector and thus introduces considerable difficulties in discrimination. We would like to limit the contribution of the zero frequency filter because the intensity of return may depend on several factors including weather, camera parameters, and look angles. On the other hand, zero frequency filters also play a role in discriminating between features. For example, the response of the zero frequency filter for water is always found to be higher than that for forest. Yet, under different viewing conditions this may be different. In order to compensate for this, we developed a scheme for normalizing the Gabor coefficients which diminishes the importance of the zero frequency component without completely discarding it. As is true with any texture representation scheme, an optimal set of filters is necessary to extract the texture information contained in the imagery.

7.4 Terrain Classification and Refinement

Classification of textural feature vectors of Gabor coefficients can be done using a number of techniques. Of them, the most prominent ones include statistical classifiers and neural networks (Kanal, 1990). While some of the techniques in the two areas share similar theories such as least-squares and clustering (e.g., k-nearest neighborhood algorithms and Kohonen's unsupervised learning) several others are entirely different and unique (e.g., Hopfield networks). The advantages that neural networks offer include their fine-grained parallelism, simple training rules, and more recently, their ability to learn arbitrary mappings (Hecht-Nielsen, 1989). In this application, we find that local evidence from the image in support of a specific feature is often not strong, due to poor contrast, low resolution, noise and obscuration from neighboring features. Neural networks have proved to excel in many such noise dominated problems and hence the choice of neural networks.

The underlying motivation for the architecture developed in this work is that subjective reasoning can only be achieved by continuous learning from examples, and not by teaching the system with an exhaustive set of rules. Indeed, it is very likely that, as we found in one of our experiments on decision-level fusion, in the process of learning the association between the texture descriptors and the feature class, the networks actually learn logical rules in fusing individual decisions from the sensors. However, such logical rules are difficult to construct when the number of target classes or the number of input feature elements are large. Furthermore, fusion of textural features from multiple sources adds to the complexity.

7.4.1 Fusion and Classification

A graceful integration of textural features computed in the form of Gabor coefficients can be achieved in two ways:

- A single vector of Gabor coefficients from corresponding cells in both radar and optical images form the input to a neural network at every cell of the image grid. Since this notion involves direct fusion of raw data features we refer to it as data level fusion (see Figure 7.3).

• Similar to many multi-level classifiers, textural feature vector from corresponding cells can be first classified independently by a lower level of neural networks, followed by fusion of their decisions with another set of neural networks at the upper level (see Figure 7.4).

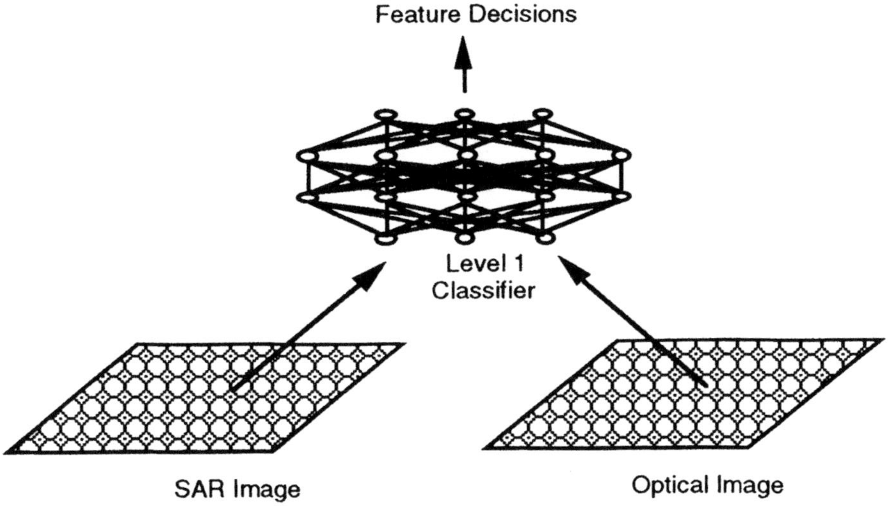

Figure 7.3 Data-level fusion of textural features

There are advantages and disadvantages with either type of fusion. The major advantage with data-level fusion is that the process of fusion takes place without any bias or loss of information about the texture behavior in the two sensors. By this we mean that the classifier is given the fullest information about the terrain feature instead of being given the biased decisions as in the case of decision-level fusion. On the other hand, the advantage with the decision-level fusion is that the input vector size at the lower level is halved (for fusion of two sensors) in camparison with data level fusion. Empirical studies into neural networks suggest that with smaller dimensionality of input vectors one could expect better performance. In addition, the learning is also faster since the number of weights decreases by the square of the input dimension.

The neural networks employed for classification are of the feedforward type and trained by the backpropagation algorithm. The choice of these networks was primarily motivated by their generalization ability and ease of interpreting the classes due to the supervised nature of the training. It is also possible to conceive other possible structures such as a hybrid combination of unsupervised Kohonen's competitive networks at the lower level and again a supervised feedforward network at the upper level to interpret the self-learned classes of Kohonen's network.

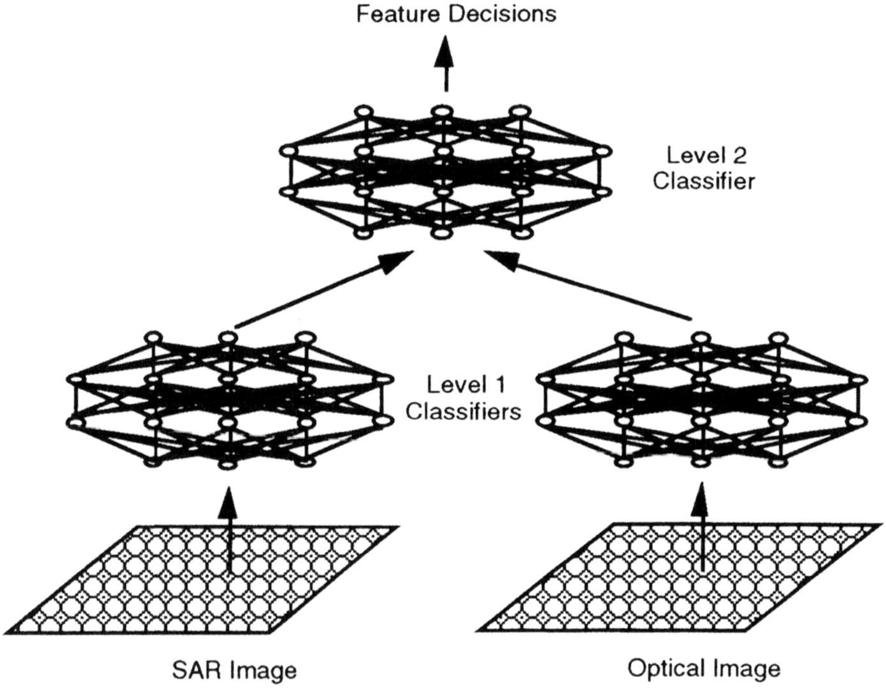

Figure 7.4 Decision-level fusion of textural features

At this point, we feel that even if such combinations are tried, the classification rate may only marginally fluctuate. Later on, as found from our experimentation, we show that while fusion of information from multiple sensors makes considerable difference in performance over individual sensors, the type of fusion architecture make relatively less difference.

7.4.2 Expert System Verification

The major role of the expert system in the InFuse architecture is to verify and validate the decisions reached by the neural networks at the second level in conjunction with the spatial database. The decisions of neural networks are represented by the respective activation values of the output neurons. Thus, because these decisions are straightforward to interpret, a sophisticated interface between the expert systems and the neural networks is not needed here. However, the other end of the expert system which communicates with the spatial database is required to have an interface to achieve a smooth flow of information in the system satisfying the database format as well as the object-oriented representation of knowledge in the expert system. This interface contains simple routines which translates queries by the expert system on the type of a terrain feature currently verified into a form that the database can directly answer. The database in our application is a Defense Mapping Agency product called Digital Feature Analysis Data (DFAD) which has polygonal structures of terrain feature information.

The expert system contains information about sensor characteristics, heuristics about region representation in imagery, and system control and management of external sources of information including the spatial database. The production rules that accomplished this utilized information from a number of sources. Domain knowledge about the sensors, including relative confidence values of their general ability to accurately classify the features of interest, is represented in production rules and in an object-oriented hierarchy within the knowledge base. Additional information about the polygonal boundaries of regions in each image resided in hand-digitized DFAD databases which were accessed initially to build the complete descriptions of each image cell, and then via the production rules as the information was needed.

The production rules may be roughly grouped according to the confidence in cell classification decisions they make. There are five such groups: (1) True Classification, (2) Assumed Classification, (3) Heuristic Classification, (4) Mixed Classification, and (5) Reject Classification. A True Classification is one that required no conflict resolution; the SAR and optical images resulted in the same classification for a given cell from each separate neural network. For example, if, for a given cell, the neural network for the SAR image converged to classify the cell as a forest, and the neural network for the optical image also converged to forest, then there is no conflict, and the expert system determines the final classification to be forest.

An Assumed Classification is one that accepts a final classification in cases where one neural network failed to converge at all. If the other converged, it was determined to be acceptable to make that classification be the final classification for the cell. A Heuristic Classification is one that relied on the observations of the experimenters in an effort to fine-tune the knowledge base. For example, it was consistently noted that when the SAR image converged to field for a given cell, and the optical image converged to water in the same cell, the true classification was actually forest. Facts of the domain such as this are easily represented in production rule format, especially since the tool we used enables rapid prototyping. In this manner, we were able to incorporate a host of heuristic rules to augment the domain.

Both Mixed Classification and Reject Classification groups refer primarily to the expert system's final classification for a given cell. If both neural networks failed to converge for a given cell, then the system accepts the failure and assigns a classification of reject to it. A cell that contains more than one feature, such as both field and water, may result in an incorrect classification or may fail to converge at all. Unless both neural networks fail to converge for a cell, conflicting classifications that are not resolved by any of the aforementioned approaches, take DFAD information into account. Since DFAD contains the polygonal boundaries of regions, it is able to determine a mixed area as one that contains the edges of more than one regional polygon. Therefore, if the DFAD database confirms that the cell is mixed, the expert system, rather than reject the cell, will classify it as being mixed.

7.5 Results and Discussion

The two approaches we adopted for natural feature extraction and classification are: (1) to extract as much information as possible using images from the different sensors individually, and then combine the results (decision level fusion), and (2) to combine information at an earlier stage, namely, at the Gabor filter response level (data level fusion).

	SAR (without Fusion)	Optical (without Fusion)	Data Level Fusion	Dec. Level Fusion	Expert Sys. Fusion
Forest	21.4 %	66.9 %	56.2 %	82.1 %	92.2 %
	65.2 %	76.4 %	86.7 %	94.3 %	92.5 %
Water	96.8 %	96.6 %	98.8 %	98.8 %	98.8 %
	96.8 %	98.3 %	99.6 %	99.6 %	98.8 %
Field	0.03 %	82.5 %	85.7 %	83.8 %	84.1 %
	0.06 %	91.7 %	94.6 %	84.2 %	92.7 %

Table 7.1 Correct Classification Rate of Natural Features in Rural Imagery

The networks can be trained on image cells at which reliable ground truth is available. We developed an interactive routine which queries the user to identify the class of a selected cell by looking at the imagery.

The results are promising with both types of approaches, although the decision level fusion performed slightly better than the data level fusion. For example, for the image named **erm27d.img** (see Figure 7.5), the correct classification rate with data level fusion is about 98%, while with decision level fusion the corresponding figure is around 94%. Table 7.1 provides the statistics for the region features in a rural area. No statistics were collected for urban images since they contain only patchy region features.

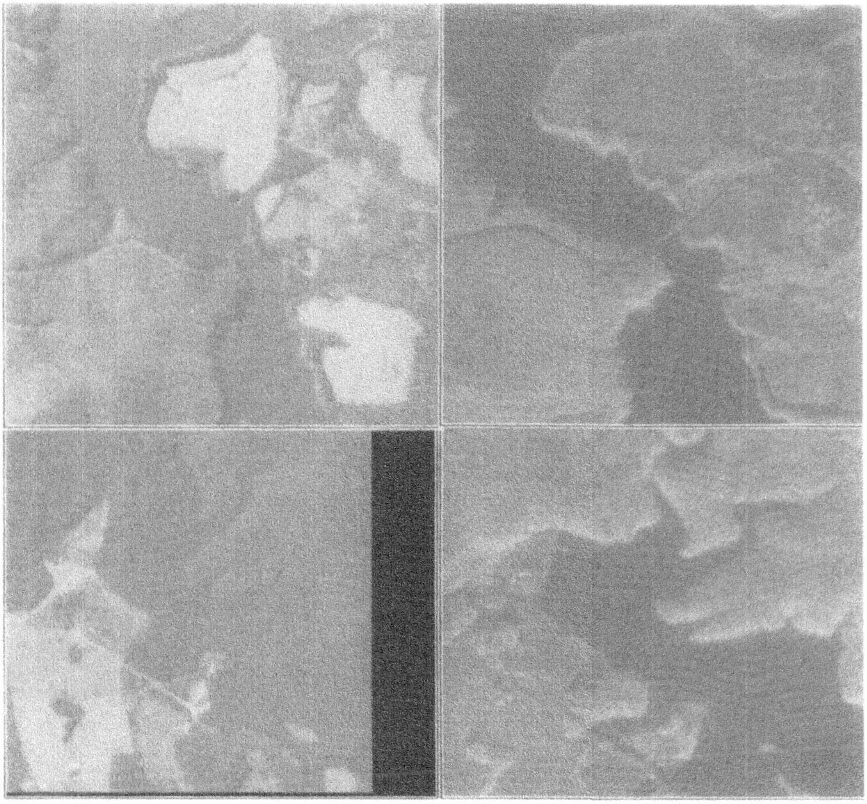

Figure 7.5 Original sample images named **eriml3.img** (top) and **erm27d.img** (bottom). The left of the pairs is the optical image and the right of the pairs is the SAR image.

The results of individual sensor nets on the two sets of rural images are presented in Figure 7.6. The sensor fusion results are shown in Figures 7.7. Recall that for the natural features, the images are divided into 16x16 cells, and each cell is classified into one of several features. When images are represented in color, a colored square is placed in the center of each cell and corresponds to the classification given to that cell. However, as shown in Figures 7.6 and 7.7, a (-) indicates water, a **square box** indicates field, a **triangle** indicates forest, and a (+) indicates an urban area. Crosses (**x**) indicate reject cells, i.e., those cells the neural networks were unable to classify. Reject cells may later be classified as mixed cells by the expert system, which is indicated by replacing the crosses with boxes.

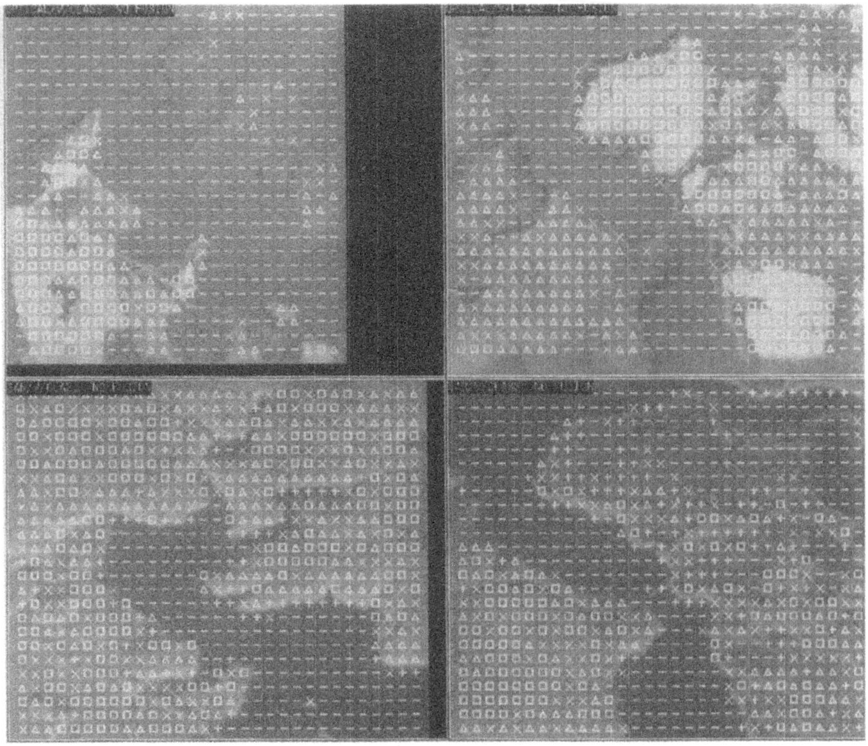

Figure 7.6 Results of classification without data fusion. See text for details.

The effect of sensor fusion is strikingly evident in these results. For instance, before applying sensor fusion, the fields in **erim13** (Figure 7.5) were classified in the SAR image as water due to dull return from the field. After the fusion (both in data level fusion as well as decision level fusion), those regions were correctly classified as fields. Similarly, in **erm27d** (Figure 7.5), there is a forest in the upper half of the image which is classified in optical imagery as water. Since in the SAR imagery, this forest is correctly classified, both types of fusion give the correct results. Some observations on these results are due.

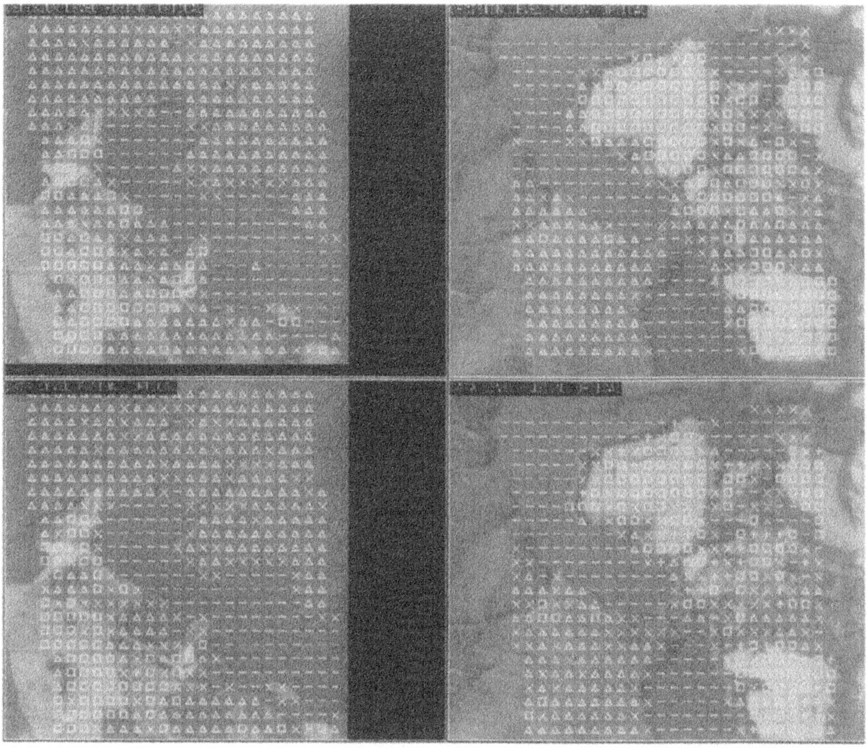

Figure 7.7 Results of classification with decision level fusion (top) and data level fusion (bottom). See text for details.

First, we point out an interesting behavior of sensor fusion at the decision level. The network was also trained with some urban regions as part of the training data. Parking lots and building roofs were erroneously classified as fields since their local texture is similar to fields in optical imagery. Clearly, humans identify them as parking lots or building roofs using the substantial knowledge we possess, such as shape and context. In order for the automated sensor fusion information to function at a higher level, it too will need to incorporate such knowledge.

A second point to mention is the dependence of rejection on the cut-off confidence level used for classification. The percentage of rejections ranged from 8% to 20% depending on this cut-off level. We investigated the effect of lowering the cut-off value on the percentage of rejection. The classification rate has been found to gracefully degrade with the cut-off almost linearly.

Rejection primarily occurred on cells in the border areas where two or more features are present. One method for handling rejection and increasing resolution of the system, is to repeat the classification process at the reject cells but with a smaller cell size. The expert system already distinguishes between those cells that, although originally rejected by both neural networks, actually are confirmed by the GIS database to be mixed. In these cells, there is more than one region that dominates. As part of the feedback from the expert system to the previous level of neural analysis, classification using smaller cell size could be specified for application only to those mixed cells.

Third, it may be noted that urban images contain several areas of strong returns in SAR. Since the strong returns tend to come from buildings and other man-made structures, it seems reasonable to conclude from these strong returns that it is an urban area. We have exploited this fact and used SAR to classify an image as urban or rural.

Fourth, an additional level of processing is necessary in order to classify an entire region as forest or water, instead of simply classifying cells. This is particularly necessary during wartime, since terrain information is usually needed in the form of regions to arrive at tactical decisions. This level of processing may be done by incorporating cooperative neural architectures to fill in the regions as much as necessary, yet retain the areas that are indicated by strong convergence of the network in that cell.

A more qualitative analysis lies in understanding how efficient the Gabor representation is and how well the network is able to generalize its decision. We have not undertaken such initiatives since the objective of this effort was to show proof-of-concept of neural networks in combination with expert systems for reliable feature extraction. The results appear to be quite promising, particularly the performance of the sensor fusion network. The results can be further improved by considering a range of ideas that resulted from this effort. For instance, a multiple resolution Gabor representation is being contemplated to facilitate validation of the features by parsing the multiple scales.

Chapter 8

Hybrid System for Multiple Target Recognition

8.1 Introduction

This chapter was written primarily by Dr. Alper Caglayan and Mr. Paul Gonzalves of Charles River Analytics, Inc. They developed a hybrid neural network/expert system approach to multiple target recognition based on their in-house hybrid development tool NueX. The work was sponsored by the Naval Surface Warfare Center, Dahlgren Division, under contract No. N60921-91-C-0186.

Multiple target recognition systems are real-time information management systems that process and assess multi-sensor data and present the best options to a decision maker. The effectiveness of current multiple target recognition systems is limited by the increasing speed of modern weapon systems and the increasing number of different types of sensors -- e.g., radar, IR, TV, EW, EO, and acoustic sensors.

The major difficulties in target detection, identification, and classification problems are the uncertainty of signal origin, the correlation of the multi-sensor data, the fluctuations in signal-noise ratio, interference from sources other than the targets intended, loss of target contact with the associated fades and gaps in the measurement data, execution of various maneuvers by the tracked targets, and the complexity of the signal processing algorithms relating the raw sensor data to target state information.

The most important characteristic of sensor data in multiple target tracking is the uncertainty of its origin. This uncertainty can arise not only from clutter, interference, and multipath effects, but also from multiple targets in the same neighborhood. Research over the last two decades has been driven to find a *computationally tractable* way of incorporating measurements of uncertain origin into existing tracks. These algorithms can be roughly divided into two groups: non-Bayesian and Bayesian. Non-Bayesian algorithms estimate target states conditioned on the correctness of their decisions on target tracks. Hence, the resulting target state estimates and covariances do not take into account the possibility of incorrect decisions. On the other hand, Bayesian algorithms for tracking problems account for the probability that the decisions may be incorrect. As discussed by Bar-Shalom (1978) and by Chang and Tabaczynski (1984), while there is no tracking algorithm capable of working reliably in an arbitrarily dense target environment, some algorithms can provide acceptable tracking performance for a limited target density.

A hybrid artificial neural network (ANN) and expert system approach can significantly enhance the performance of current multiple target tracking systems. Artificial neural networks produce a nearest neighbor classifier. Conceptually, artificial neural networks are applicable to multiple target tracking problems by learning the spatiotemporal attributes of target trajectories and classifying multi-sensor data. For instance, a multilayered perceptron has been trained to model optimum guidance trajectories for the aeroassisted orbital plane change scenario in Caglayan and Allen (1990).

The integration of neural networks, expert systems, and analytic algorithms is particularly suited for the development of intelligent MTR systems. The objective of our hybrid approach is to take advantage of the complementary strengths of each approach while negating individual weaknesses. For instance, neural networks offer robust pattern recognition performance, fault tolerance, and parallel processing capabilities but have difficulty with hard constraints. Expert systems offer symbolic processing capability when the domain knowledge can be structured into rules but have difficulty with incomplete information. Analytic algorithms offer excellent performance when there are no significant modeling errors while having degraded performance under modeling errors.

The use of neural networks in scan-scan correlation in multi-target tracking has been investigated by Kuczewski (1989). A discussion of the potential impact that neural network technology may have on target recognition systems is found in Roth (1990). An artificial neural network approach to multiple target tracking offers several advantages including a rapidly adaptable on-line solution via new models and learning algorithms (e.g., addressing the need for adaptation to target and environment changes), implementation efficiency on the emerging neural computers, and off-line productivity improvement by reducing trial and error.

While neural networks can play a significant role in multiple target tracking, a pure neural net approach would mimic only the right hemisphere functions of the human brain. In order to incorporate formal knowledge (i.e., probabilistic algorithms) and qualitative reasoning of human experts, an expert system approach needs to be added, emulating the left hemisphere function of the human brain.

8.1.1 Objectives and Approach

The major objective of our study was to demonstrate the feasibility of a hybrid neural network/expert system approach to multiple target recognition (MTR). To accomplish this objective, we performed the following tasks:

- Specification of a hybrid MTR system architecture incorporating conventional signal processing and probabilistic tracking algorithms with ANN and Knowledge Base (KB) classifiers, and a KB decision maker.
- Design and implementation of two competing hybrid MTR systems using the developed architecture: one with invariance-based ANN input features and another with ANN features generated by a conventional tracking estimator.
- Through trade studies, comparison of the performance of the two competing hybrid MTR systems. In addition, we compare the individual performance of the ANN classifier, the KB classifier, and the hybrid classifier for each system.

The general technical approach to developing a hybrid neural network/knowledge based multiple target recognition system and demonstrating feasibility by prototype demonstration centers on 1) the determination of a hybrid architecture which will use both the on-line classification and parallel processing of neural networks and the formal knowledge and reasoning of domain experts; 2) the design of a hybrid system using the defined architecture; 3) the development of a prototype system to demonstrate feasibility; 4) evaluation of the performance of the prototype system via trade studies; and 5) determination of requirements and objectives for a full-scale research prototype for real-time demonstration.

The first step in developing a hybrid multiple target recognition system is the specification of the architecture to be employed. This involves determining the neural network, knowledge base, and conventional signal processing/estimation theoretic components of the architecture and specifying the topology of the system vis-a-vis how the various neural networks, knowledge bases, and conventional subsystems are interconnected and how they interface with each other. Specific tasks are assigned to elements within the system such as sensor data fusion, feature extraction from sensor data, varying levels of object classification, hypothesis generation, and high level decision-making. Suitable tasks for existing analytical techniques such as probabilistic multi-target tracking algorithms are also determined.

The second stage of the effort entails the design of a hybrid system. Specifically for neural networks, this stage involves the selection of the neural network paradigm, network topology, and the creation of training and evaluation databases to ensure efficient training performance with generalization capability. On the expert system side, domain specific knowledge representation is encoded within the various knowledge bases and also evaluated to ensure optimal reasoning performance.

The third stage involves the implementation of the hybrid multi-target recognition system into a prototype system. This is accomplished via our in-house hybrid neural network/expert system software development tool, NueX. The HyperCard based NueX provides for graphical neural network specification, editing and training of networks, dialog box rule specification, and a forward-chaining inference engine. Applications are developed as stacks so that the application graphical user interface can be easily developed using the HyperCard programmable objects, and the conventional signal processing and estimation theoretic algorithms can be programmed using the HyperTalk scripting language.

The fourth step in the research effort involves using the developed prototype system and evaluating its performance via trade studies. Included in these trade studies are how well does the system perform for various sensor noise levels, or under varying object densities. Specific metrics employed include probability of missed detection, probability of false alarm, and detection time.

The fifth and final stage of the effort is the formulation of requirements and a full-scale research prototype for Phase II development and real-time demonstration. Hardware system issues, including neural computer implementation, would be addressed.

8.1.2 Summary and Results

Our developed hybrid MTR architecture employs a hierarchical structure. The highest level, represented by the executive knowledge base, performs the overall decision-making, management, and coordination functions for the system. This includes data input/output management, coordination with the lower level, and overall target classification. The lower level consists of neural network classifiers, knowledge base classifiers, and analytic algorithms for ANN input feature and KB classifier input generation. The results of the ANNs and the classification knowledge base are fed back to the executive knowledge base for higher-level decision-making.

Two hybrid systems have been implemented using our in-house hybrid neural network/expert system development tool, NueX. The first hybrid systems consists of:

- a target tracking estimator which computes radar measurement residuals assuming the target is executing a hostile guidance law
- an ANN classifier driven by the radar measurement residuals
- a knowledge base (KB) classifier processing the radar measurements directly

- a knowledge base decision maker integrating the ANN and KB classification decisions

The second hybrid system consists of:

- a sensor feature extraction algorithm based on invariance
- an ANN classifier driven by the sensor features
- a knowledge base (KB) classifier processing the radar measurements directly
- a knowledge base decision maker integrating the ANN and KB classification decisions

The first hybrid system's performance is primarily influenced by the tracking algorithm error under modeling errors, typical of probabilistic tracking algorithms where the target acceleration is modeled as filtered white noise. On the other hand, the second hybrid system uses a more direct neural network approach where the analytic algorithms are used only for feature extraction without making *any* modeling assumptions about the target dynamics.

Trade studies were done to evaluate the performance of the prototype hybrid MTR system. In particular, two trade studies were performed covering:

- variation in radar sensor signal-to-noise ratio (SNR)
- variation in target density.

For each trade study, the performance of the residual-based and the invariance-based feature hybrid systems are compared. In addition, the performance of the individual elements of the hybrid system were evaluated, i.e., the performance of the ANN classifiers, the knowledge base classifier, and the hybrid classifier. Performance metrics include probability of missed detection, probability of false alarm, and average detection time for hostiles.

In general, our demonstration study shows the following:

- A hybrid neural network expert system approach to MTR incorporating conventional signal processing and tracking estimators is feasible.
- The performance of the hybrid classifier is higher than the performance of the individual ANN and KB classifiers comprising the hybrid system.
- The invariance-based sensor feature driven ANN outperforms the tracking estimator residuals driven ANN.

In particular the trade studies show the following:

- The invariance-based MTR outperforms the residuals-based MTR across radar SNR variation of 40-15 dB. For the invariance based MTR, probability of missed detection rate is under 10%, the false alarm rate is around 10%, and the detection time is 2 seconds. In contrast, probability of missed detection is 2%, the false alarm rate is 25-40%, and the detection time is 6 seconds for the residual-based MTR. The poor false alarm performance of the residuals-based MTR is due to the modeling errors in the tracking estimator. The missed detection performance of the hybrid classifier is better than either the ANN and KB classifier performance. In contrast, the false alarm performance of the hybrid MTR is higher than the ANN or KB classifier performances.

- Under object density variation between 25, 50, 75, and 100 targets, for the invariance-based MTR, probability of missed detection remains under 5% while no appreciable effect is noticeable for false alarm rates. For the residual-based MTR, missed detections show no trend with increased density, but false alarms decrease at the higher densities partly due to delays as the number of targets increase. Average detection times for hostiles increase with density level for both hybrid MTR systems with once again the invariance feature MTR system outperforming the residual-based MTR system by a factor of 3 to 1.

8.2 Hybrid System Description

8.2.1 Hybrid Neural Network Expert Systems

The integration of neural networks and knowledge based expert systems is an ideal step in the development of intelligent systems. In general, neural networks provide *soft* constraints, while expert systems allow *hard* constraints (Glover and Rao, 1990). The two methods complement each other in a manner similar to the hemispheres which contain the human mind. Neural networks, corresponding to the right hemisphere of the brain (Ruch, 1984), consist of nonlinear functions, pattern recognition capabilities, fault tolerance and parallel processing; expert systems, which relate to the left hemisphere, involve sequential calculations, language processing, formal logic and rule interpretation. The objective of a hybrid approach is to take advantage of the complementary strengths and weaknesses of each method to create a system that can outperform either method alone.

Topologies for Hybrid Systems:

Several potential architectures are possible for implementing hybrid intelligent applications incorporating both Artificial Neural Networks (ANN) and Expert Systems (ES). On the most simplistic end are either the parallel or sequential connections of a hybrid system consisting of one ANN and one ES. A system can have parallel connections with system inputs simultaneously being fed to the ES and ANN and outputs then being collected. The parallel architecture offers a comparative or cooperative hybrid system scheme; that is, outputs of the ES and ANN can be compared or used in conjunction to generate system outputs. In sequential connections, the first element can act as a feature extractor from system inputs and generate the required feature inputs for the next element. Other simple parallel and sequential connections exist including two expert systems or two ANNs. As a target recognition example, a parallel connection is possible with one network taking system inputs (e.g., various sensor measurements from a sensor suite) and performing feature extraction to provide the required input features for another network.

More intricate hybrid system topologies are derived from these simple subsystems. Such intricate topologies can be derived from the blackboard paradigm and from hierarchical architectures. The hybrid Multi-Target Recognition (MTR) system developed in this study implements a form of connection scheme employing a hierarchical architecture.

Blackboard Paradigm:

The blackboard paradigm consists of a mechanism on which the problem and initial data are written and a team of specialists each looking for an opportunity to contribute to the development of the solution. Blackboard systems have been used in applications problem solving requiring heterogeneous representations. For instance, the Expert Navigator (Glasson and Pomarede , 1986) uses a blackboard system. The blackboard based problem solving is ideally suited to the multi-target recognition system where the expertise can be modularized into trajectory estimation, feature extraction, and object identification.

The blackboard system models are composed of independent knowledge sources, diverse knowledge representation and reference mechanisms, a blackboard which is a global database containing data and partial solutions, and a control component which makes the high-level decisions about the problem-solving and resource allocation. In our proposed hybrid neural network/expert system environment, knowledge sources can be implemented using knowledge bases or neural networks. The blackboard can be implemented as a global variable object within a HyperCard script. The controller can be implemented using a knowledge base. The main advantage of our hybrid approach is that all of these different knowledge representation constructs will be included in the environment. Figure 8.1 shows a possible blackboard architecture for a MTR system.

Hierarchical Systems:

Figure 8.2 illustrates a simplified functional architecture for the MTR system developed and used in this study. The architecture follows that seen in the literature on hierarchical and intelligent control systems (e.g., Antsaklis, Passino and Wang, 1991) and Acar and Ozguner, 1990).

In intelligent control, the top level, or the organizational level, provides the interface to the human operator and the overall management, decision-making, and learning for the system. The middle level, or the coordination level, provides the link between the organization level and the execution level. It also involves management, decision-making, and learning but at the more detailed and specific subsystem level. The third and final level, the execution level, provides the interface to actual subsystem hardware (sensors and actuators) and the external environment.

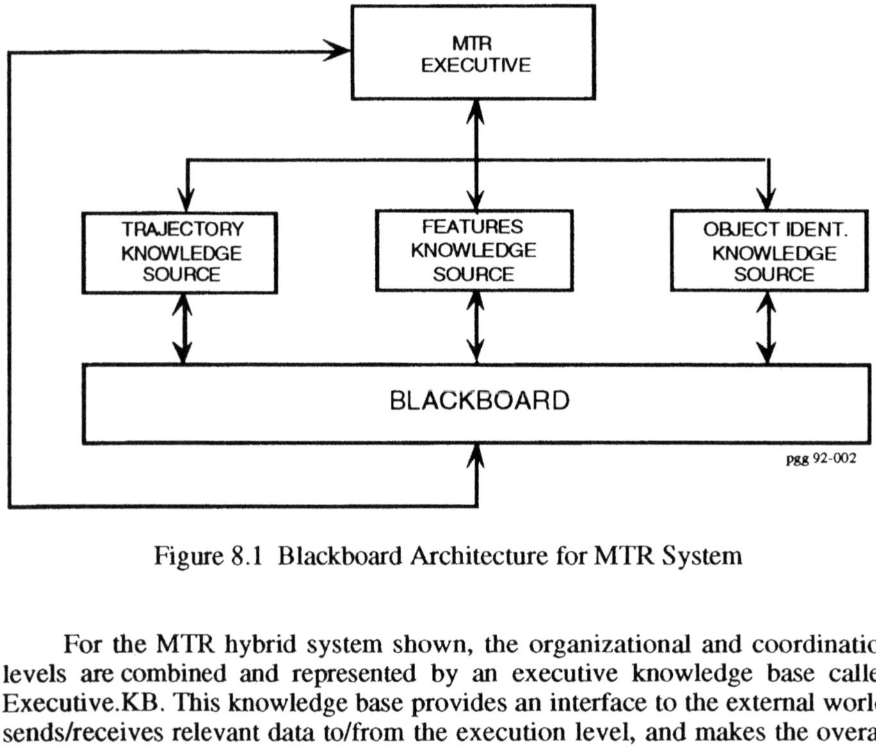

Figure 8.1 Blackboard Architecture for MTR System

For the MTR hybrid system shown, the organizational and coordination levels are combined and represented by an executive knowledge base called Executive.KB. This knowledge base provides an interface to the external world, sends/receives relevant data to/from the execution level, and makes the overall decision and object identity (friendly or foe). The execution level consists of two neural networks and a classification knowledge base grouped in a parallel manner. The two neural networks employ different input features (scaled estimator velocity residual features for one and an invariant-based feature set for the other) and output a determination of object identity. The detection knowledge base has the same type of output as the ANNs but is based on a different set of input features. The results of the ANNs and the classification knowledge base are fed back to the executive knowledge base for higher-level decision-making.

Several aspects of this hierarchical structure are evident:

- The higher levels are concerned with broader aspects and overall system behavior, whereas the lower levels are concerned with more localized aspects. For example, the organizational/coordination level may detect an unknown object within sensor range. It would then interrogate the execution level to make a determination, if possible, of object identity.

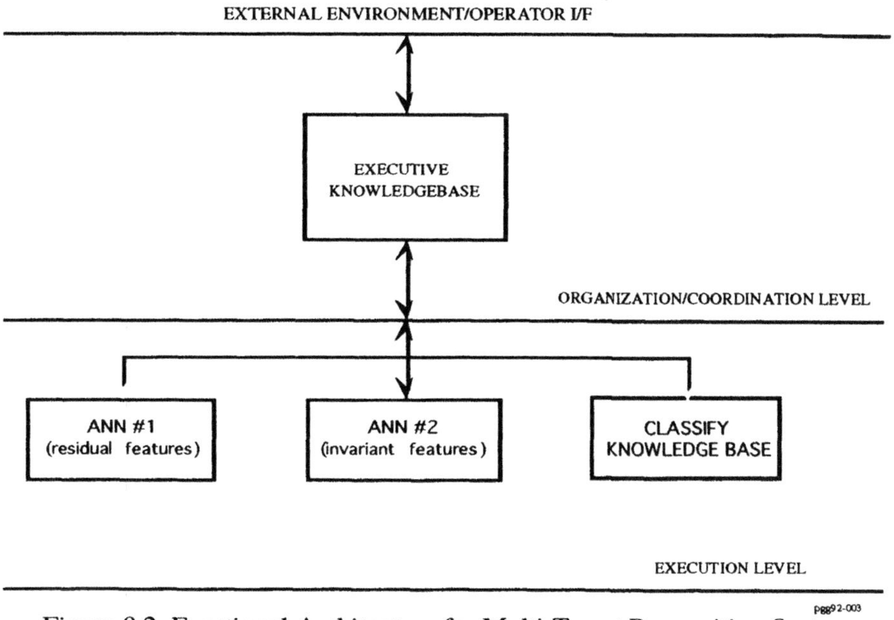

Figure 8.2 Functional Architecture for Multi-Target Recognition System

- The structure employs the "principle of increasing intelligence with decreasing precision" (Saridis, 1989). That is, the amount of information content increases at higher levels, while the precision of the information decreases. This is reflected in more abstract models being used at higher levels with an increase in model granularity at the lower levels.

8.2.2 Hybrid System Development Tool-NueX

NueX provides an application development system incorporating both neural networks and expert systems within the functionality of the Apple Macintosh HyperCard environment. The *NueX* hybrid environment is illustrated in figure 8.3. The environment is divided into three main sections: Neural Networks, Expert Systems, and Hybrid. Typically, the development process begins with the ANN and knowledge base development. After developing the ANN architectures and setting up the KB rules, the links between ANN input/output nodes and specific KB objects/properties are defined. ANN training is accomplished either interactively or automatically by using the expert neural training, which performs the learning supervision. ANNs and KBs are then tested within their respective sections, and design iterations are performed as needed. Within the Hybrid section, the specific application user interface is created by taking advantage of HyperCard's object-oriented visual programming language in addition to predefined graphical and data I/O routines.

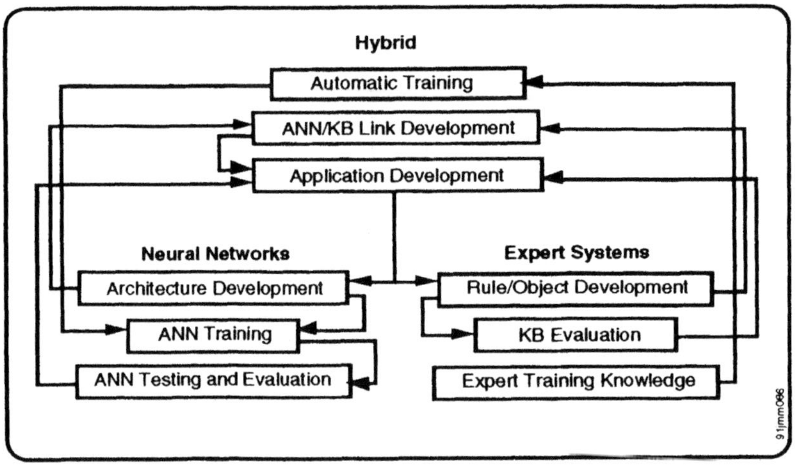

Figure 8.3 *NueX* Hybrid Development Environment

Neural Network Architecture Development

The ANN Architecture Development section of the hybrid *NueX* environment is illustrated in figure 8.4. Shown graphically is the classic XOR network (Rumelhard, 1988) with its associated two input nodes, two hidden nodes, and a single output node. The palette to the left of the development window contains the tools necessary for specifying the ANN interconnect structure. Using this palette, the user can move node (the currently selected option in figure 8.4), create and erase nodes and paths, and display information separately for each path and node.

NueX also provides automatic ANN development routines which can be incorporated into specific applications. These routines can either create user-specified networks as required or choose from a set of predefined architectures. Fully connected architectures can easily be specified between user-defined layers of nodes. Complete ANN architectures can also be specified through a series of prompts which request the number of layers and nodes/layer.

Neural Network Training and Evaluation

In the interactive training option, the user is required to monitor and control all aspects of the training process. The following functions are available:

Deviate Weights - deviate weights randomly or by user-specified amount
Reset Weights - reset all weights randomly with specified range
Cycle Order - user-defined layer propagation cycle order
Graph Freq. - specify graphing frequency (cycles/graph update)
Learning Cycles - specify number of learning cycle iterations
Randomize IOPs - automatically randomize training patterns
Start/Stop Recording - save RMS error plotting data to a text file
Use Biases - set the network's bias output between 0 and 1

Figure 8.5 shows the graphical interface for training and evaluating neural networks. The current RMS training error is plotted using a logarithmic scale; this serves to visually amplify the RMS signal as the network approaches convergence. The current RMS error is continually updated graphically and within the text fields as the learning progresses. The ANN's current learning rate, momentum, and bias output are also displayed within text fields.

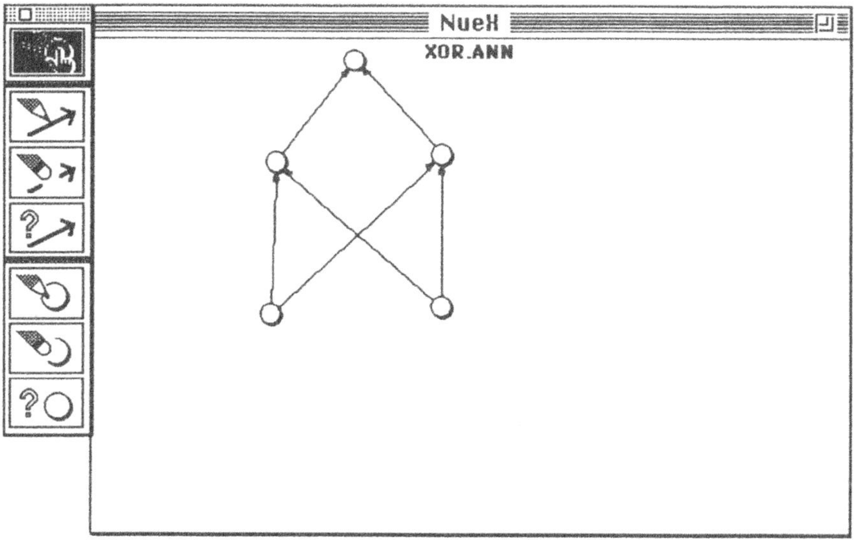

Figure 8.4 ANN Architecture Development

Knowledge Base Development and Evaluation

NueX provides the functionality to design, develop, and evaluate expert system knowledge bases. This functionality is provided with a Knowledge Base editor for developing rule base, and a knowledge base monitor to test and evaluate the inference performance of the developed rule base. Shown in figure

8.6 is the KB editor interface used to create rules. The rule shown here is from a target classifier knowledge base used in our prototype hybrid MTR system.

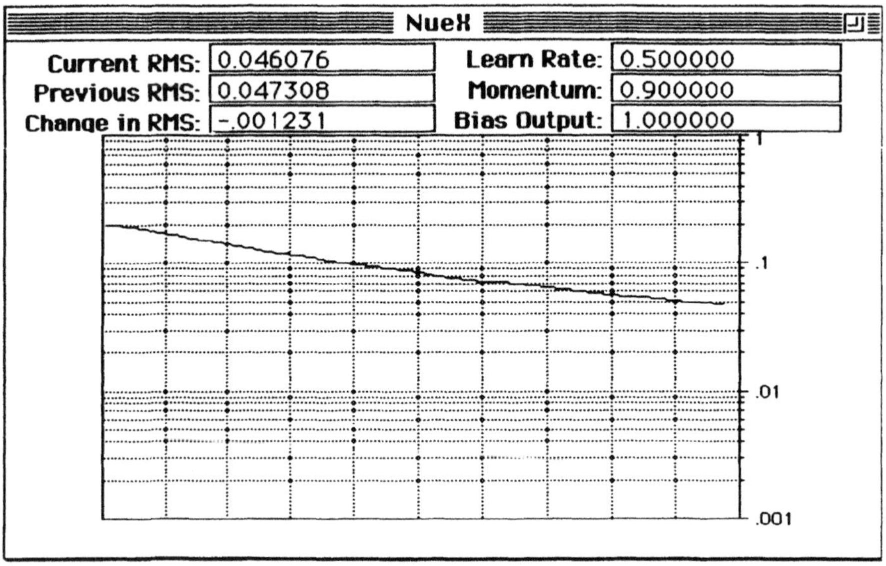

Figure 8.5 ANN Training and Evaluation

The elements of the editor that pertain to rule generation include, rule name, rule firing priority, the IF or conditional part, and the THEN part or the facts that get asserted when the rule is fired. The rule name is the symbolic name the user can assign to the specific rules to ease in the knowledge base development and evaluation process. The rule firing priority is an integer value greater than or equal to zero. The higher the integer, the earlier the rule will fire. The conditional part of the rule is given in the text window region under the IF in figure 8.6. Logical expressions are input here by the user. The example shown here looks at current values of range, range rate, and cosine of Target Aspect Angle (TAA) to see if certain conditions are met. The user can select whether to employ the logical AND or the logical OR when evaluating the conditional. This is done by checking the boxes next to AND OR, respectively, as shown in figure 8.6.

The facts that are asserted when the rule is fired are in the text window below the THEN. Elements of this text window are scripted using a basic knowledge of HyperCard's scripting language HyperTalk. For instance, in the rule shown if the IF part is true the fact asserted is that KB_output is "hostile." Rules can be reset at the end of an inference cycle by checking the box next to

RESET. NueX also affords other capabilities for knowledge base development including, definition of global variables, use of default values for global variables, cutting and pasting rules, and capabilities for accessing and calling ANNs and other knowledge bases.

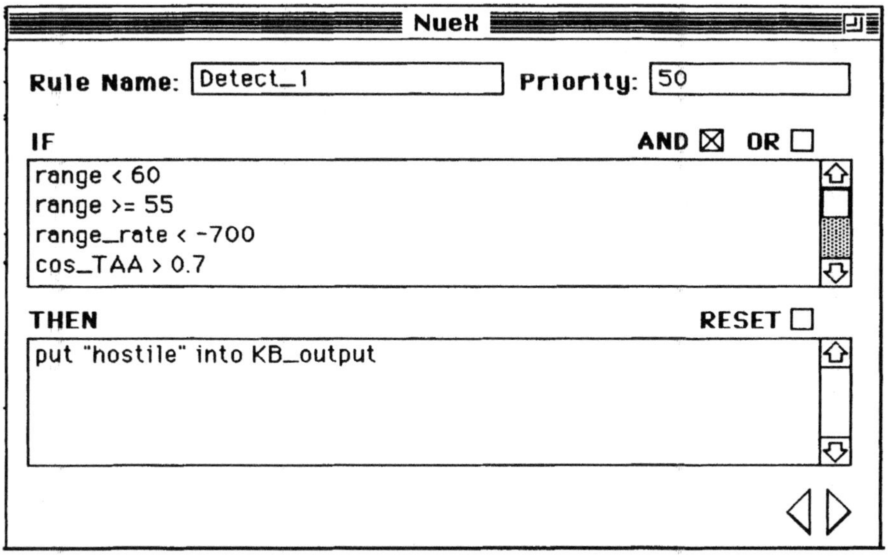

Figure 8.6 Knowledge Base Development

Knowledge bases can be tested and evaluated using the KB monitor shown in figure 8.7. The example shown here is another knowledge base (Executive.KB) from the MTR system. Shown on the left-hand side are the thirteen rules in the knowledge base named Exec_1 through Exec_13. Also shown to the right of the rule names are all the global variables along with their current values and their default values if any. Default values are the values assigned to global variable when the knowledge base is invoked. The user can test and evaluate the knowledge base by either running through the whole inferencing process or stepping through one rule at a time. While running the knowledge base, the current invoked rule is shown highlighted and any rules fired are listed in order of firing in the column to the left of the rule names. During the knowledge base run through, the values for the global variables are updated on the screen. The step through is similar except that the inferencing stops when a rule is fired. The user can check the global variables before stepping through until the next rule fires. In a such a manner, a knowledge base can be debugged, tested, and evaluated.

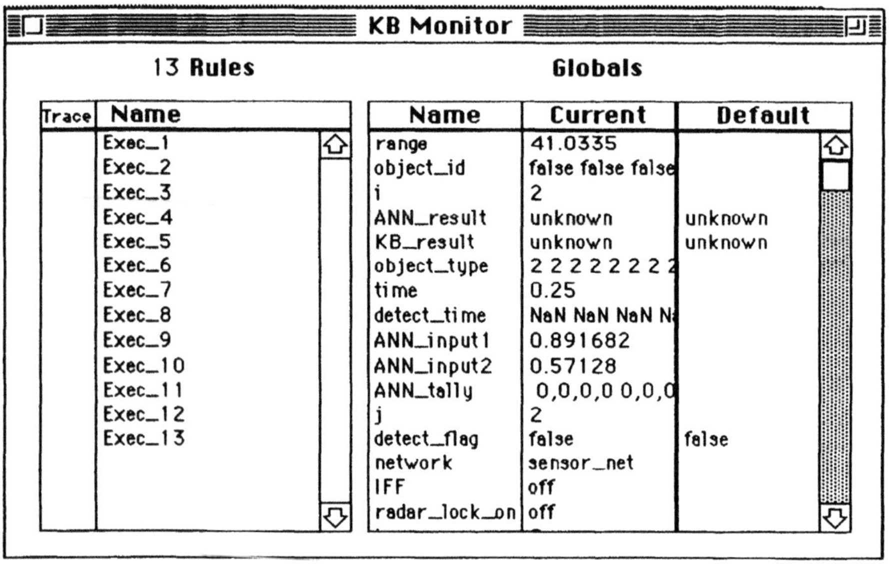

Figure 8.7 Knowledge Base Evaluation

8.3 Hybrid System Implementation

This section describes the implementation of the hybrid neural network/expert system for multiple target recognition.

8.3.1 Simulation Overview

Figure 8.8 presents an overview block diagram of the hybrid MTR system implementation. The implementation is broken down into two specific areas: the external modules that provide off-line generation of trajectory sensor data and the MTR hybrid system. The external modules include the following blocks:

- target and ownship models that are used to generate object/ownship position and velocity states
- a radar sensor subsystem block that generates measurements of target relative states

Two hybrid architectures have been implemented. The first hybrid system consists of:

- a target tracking estimator which computes radar measurement residuals assuming the target is executing a hostile guidance law
- an ANN classifier driven by the measurement residuals
- a knowledge base (KB) classifier processing the sensor measurements directly

- a knowledge base decision maker integrating the ANN and KB classification decisions

The second hybrid system consists of:

- a sensor feature extraction algorithm based on invariance
- an ANN classifier driven by the sensor features
- a knowledge base (KB) classifier processing the sensor measurements directly
- a knowledge base decision maker integrating the ANN and KB classification decisions

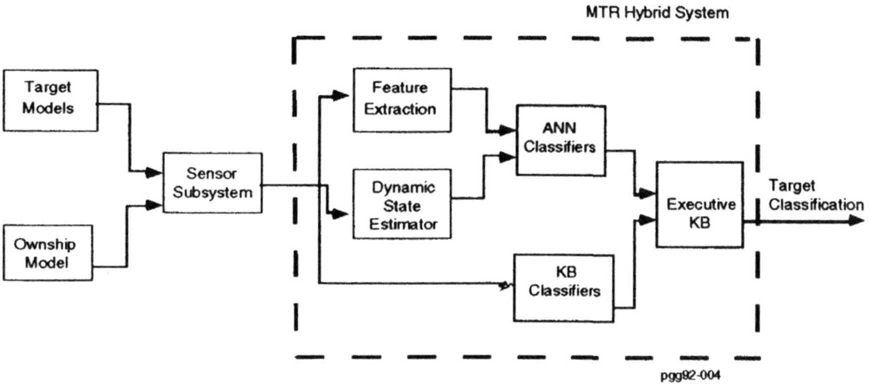

Figure 8.8 Overview Block Diagram of MTR System Implementation

The first hybrid system's performance is primarily influenced by the tracking algorithm error under modeling errors, typical of probabilistic tracking algorithms. On the other hand, the second hybrid system uses a more direct neural network approach where the analytic algorithms are used only for feature extraction without making *any* modeling assumptions about the target dynamics. The system processes the data to classify objects as to target potential. Implementation of the MTR system has been with our in-house hybrid development tool NueX. Functional descriptions of the external modules are provided in section 8.3.2, whereas section 8.3.3 provides descriptions of the MTR hybrid system modules.

8.3.2 External Modules

The Target and Ownship modules simulate multiple hostile and friendly targets and the tracking platform. The modules generate time histories of target and platform position, velocity, and acceleration. The modules currently support three modes of trajectory generation operation: 1) trajectories employing a turn-to-heading intercept guidance; 2) trajectories employing a line-of-sight (LOS) intercept; and 3) trajectories using zero acceleration/constant velocity. In general, the modules implement the following vector differential equations:

$$\dot{\mathbf{r}}(t) = \mathbf{v}(t)$$
$$\dot{\mathbf{v}}(t) = \mathbf{a}(t) \tag{8.1}$$

where $\mathbf{r}(t)$ is the target's position at time t, $\mathbf{v}(t)$ its velocity, and $\mathbf{a}(t)$ its acceleration. The acceleration is defined for three different guidance schemes in the following manner. For constant velocity it is given by:

$$\mathbf{a}(t) = 0 \tag{8.2a}$$

For the LOS intercept it is:

$$\mathbf{a}(t) = - K \dot{\rho} (\mathbf{u}_r \times \Omega) \tag{8.2b}$$

where $\dot{\rho}$ is range rate, $\mathbf{u}_r$ is the unit LOS vector to the target, Ω is the rotation of the LOS vector, and K is a gain parameter usually chosen to be 3 or 4 or greater depending on vehicle maneuverability. The turn-to-heading intercept results in a zero z component of $\mathbf{a}(t)$ and with the following x and y components:

$$a_x(t) = - |\mathbf{v}| \dot{\psi} \sin(\psi)$$
$$a_y(t) = |\mathbf{v}| \dot{\psi} \cos(\psi) \tag{8.2c}$$

where $\mathbf{v}$ is the current velocity and the heading rate command ($\dot{\Psi}$) is given by:

$$\dot{\psi} = \text{signum}\{\psi_d - \psi\} \cdot \min\{|\psi_d - \psi|, T_r\} \tag{8.3}$$

with Ψ_d the desired heading (i.e., the heading that points directly to the target), Ψ the current heading, and T_r the maximum allowable turn rate.

Figure 8.9 shows the Radar Sensor Module block diagram model of the MTR radar sensor subsystem. The MTR sensor path is driven by the target and ownship position and velocity vectors, $\mathbf{r}_t$, $\mathbf{v}_t$, and $\mathbf{r}$, $\mathbf{v}$, respectively. Their difference yields an ownship relative set of position and velocity states ($\mathbf{r}_r$ and $\mathbf{v}_r$). These relative states are then transformed from cartesian into spherical coordinates. The resulting six dimensional spherical states (range (ρ), elevation (θ), azimuth (Ψ), range rate ($\dot{\rho}$), elevation rate ($\dot{\theta}$), and azimuth rate ($\dot{\Psi}$) are then corrupted with noise to simulate radar sensor measurements. The range and range rate signals are corrupted via multiplicative noise, i.e.,

$$\tilde{\rho}(t) = \rho(t) + k\,\rho(t)\,w_\rho(t)$$
$$\tilde{\dot{\rho}}(t) = \dot{\rho}(t) + k\,\dot{\rho}(t)\,w_{\dot{\rho}}(t) \tag{8.4}$$

where the ~ denotes the noisy measurement, k is the multiplicative factor, and **w** are the white noise sources. The angles and angular rates are corrupted with additive noise in the following manner:

$$\tilde{\theta}(t) = \theta(t) + w_\theta(t)$$
$$\tilde{\psi}(t) = \psi(t) + w_\psi(t)$$
$$\tilde{\dot{\theta}}(t) = \dot{\theta}(t) + w_{\dot{\theta}}(t)$$
$$\tilde{\dot{\psi}}(t) = \dot{\psi}(t) + w_{\dot{\psi}}(t) \tag{8.5}$$

with again the ~ denotes the noisy measurements and **w** are the white noise sources.

The six independent noise sources are uncorrelated sequences with zero-mean normal distribution. For simplicity, covariances are chosen to yield +/-1 degree errors in angular measurements and +/-.1 degree/s errors in angular rates. For the range and range rate measurements, the unit variance noise source are multiplied by the current range/range rate and by a multiplicative factor, k. The values chosen correspond to a signal-to-noise ratio (SNR) of 25 db.

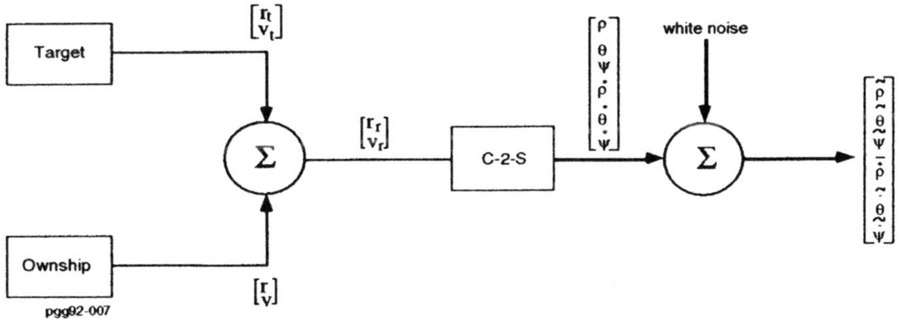

Figure 8.9 Sensor Subsystem Model

8.3.3 MTR Hybrid System

The MTR hybrid system provides on-line classification of targets based on sensor measurements. It uses both knowledge bases and neural networks to perform the classification function. A functional block diagram of the hybrid system is shown in figure 8.10. As shown, an executive knowledge base (Executive.KB) controls the overall classification process by interrogating in parallel one of two neural networks (Resid_features.ANN or Sensor_features.ANN) and another knowledge base (Classifier.KB). The user specifies which of the two networks to use. The two networks have the same topology but use different input features, i.e., sensor-derived invariance-based features or estimator generated velocity measurement residuals. Results of the neural network and the classification knowledge base are integrated by the executive knowledge base, which then makes the final decision on target classification. In the rest of this section we further describe the individual elements of MTR Hybrid System and the implementation of the system within our in-house hybrid development too NueX.

The Dynamic State Estimator is based on the extended Kalman Filter algorithm. As general background, consider the linear discrete-time system described by

$$\mathbf{x}_{k+1} = \mathbf{A}\mathbf{x}_k + \mathbf{B}\mathbf{x}_k + \mathbf{E}\mathbf{w}_k \tag{8.6}$$

where $\mathbf{x}$ is a vector of system states, $\mathbf{u}$ the controls, and where $\mathbf{w}$ the process noise is a vector of random sequences with zero mean, no time correlation (i.e. *white* noise), and with covariance matrix $\mathbf{Q}$. At each time step k, a measurement $\mathbf{z}_k$ is made, which is related to the state $\mathbf{x}_k$ by

$$\mathbf{z}_k = \mathbf{C}\mathbf{x}_k + \mathbf{v}_k \tag{8.7}$$

where $\mathbf{v}$ is *white* measurement noise with zero mean and a covariance matrix $\mathbf{R}$. Given the noisy measurement $\mathbf{z}_k$, the extended Kalman Filter (EKF) algorithm (Bryson and Ho, 1975) computes the least mean sequence estimate of the state via the following equations

$$\hat{\mathbf{x}}_{k+1} = \overline{\mathbf{x}}_{k+1} + \mathbf{K}_{k+1}\left[\mathbf{z}_{k+1} - \mathbf{C}\overline{\mathbf{x}}_{k+1}\right] \tag{8.8}$$

$$\mathbf{P}_{k+1} = (\mathbf{I} - \mathbf{K}_{k+1}\,\mathbf{C})\;\mathbf{M}_{k+1} \tag{8.9}$$

and the a priori state estimate and error covariance matrix propagated as

$$\overline{\mathbf{x}}_{k+1} = \mathbf{A}\hat{\mathbf{x}}_k + \mathbf{B}\mathbf{u}_k \tag{8.10}$$

$$\overline{\mathbf{M}}_{k+1} = \mathbf{AP}_k\mathbf{A}^T + \mathbf{EQE}^T \tag{8.11}$$

with the filter gain $\mathbf{K}_k$ given by

$$\mathbf{K}_{k+1} = \mathbf{M}_{k+1}\,\mathbf{C}^T\left(\mathbf{CM}_{k+1}\,\mathbf{C}^T + \mathbf{R}_{k+1}\right) \tag{8.12}$$

We now apply the extended Kalman Filter formulation to the estimation problem at hand. We use a relative coordinate system in which only the target is accelerating. The system dynamics model is then given by

$$\begin{aligned} \dot{\mathbf{r}} &= \mathbf{v} \\ \dot{\mathbf{v}} &= \mathbf{a}(t) + w \end{aligned} \tag{8.13}$$

where **r** and **v** are the relative three-dimensional cartesian position and velocity state vectors between the target and the ownship. The input term **a**(t) is the relative acceleration vector. If ownship is stationary or non-accelerating, **a**(t) becomes the target's acceleration. In order to drive the Kalman Filter, we make the *assumption* that all targets are using a turn-to-heading intercept guidance law in which **a**(t) is given by (8.2c). Therefore, at each time step, and for each

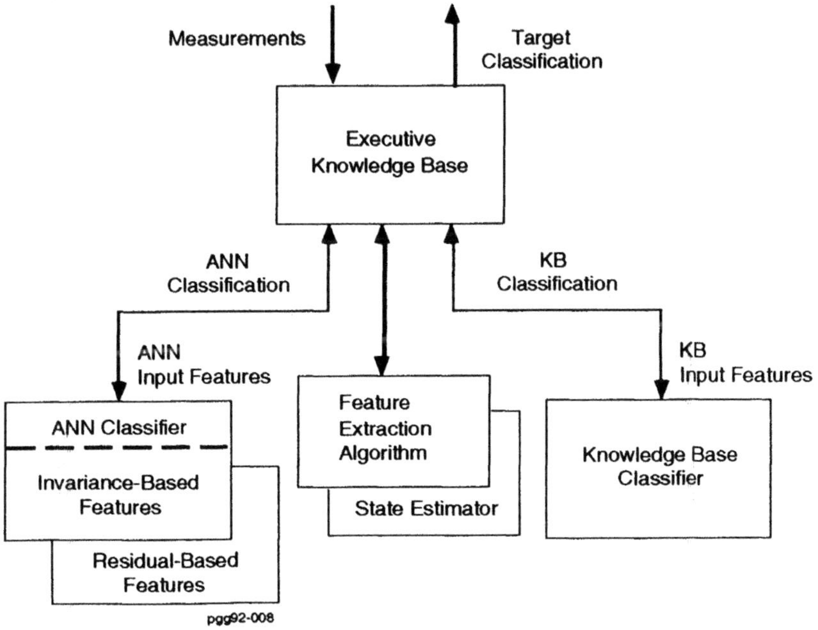

Figure 8.10 Functional Block Diagram of the Hybrid System

target, **a**(t) is calculated via (8.2c) using current state estimates and then fed to the state estimator to generate new estimates of target relative position and velocity. Hence, the measurement residuals of this tracking filter will be zero-mean when the target is hostile and executing the assumed intercept law. In contrast, the measurement residuals will be non-zero mean when the target is friendly. Noisy sensor measurements of target range, elevation, azimuth, and associated rates are made at each time step and are as given by (8.4) and (8.5).

The Feature Extractor allows for the reduction and transformation of the data input set into pertinent features that the ANNs can use for classification purposes. For the problem at hand the data input set consists of sensor-derived measurements of position and velocity in spherical coordinates, and filter generated state estimates of position and velocity in cartesian coordinates. From our analysis, two neural network input features were derived:

- estimator generated target velocity measurement residuals and
- sensor derived target features based on invariance.

The first set of features is based on relative velocity measurement residuals in the self-centered cartesian frame of reference given by

$$\tilde{v}_x(k) = v_x(k) - \hat{v}_x(k|k-1)$$
$$\tilde{v}_y(k) = v_y(k) - \hat{v}_y(k|k-1) \qquad (8.14)$$

where v_x is the x-velocity pseudomeasurement, $\hat{v}(k|k-1)$ is the single stage prediction of the x-velocity pseudomeasurement generated by the target state estimator described in the previous section and $\tilde{v}_x(k)$ is the corresponding

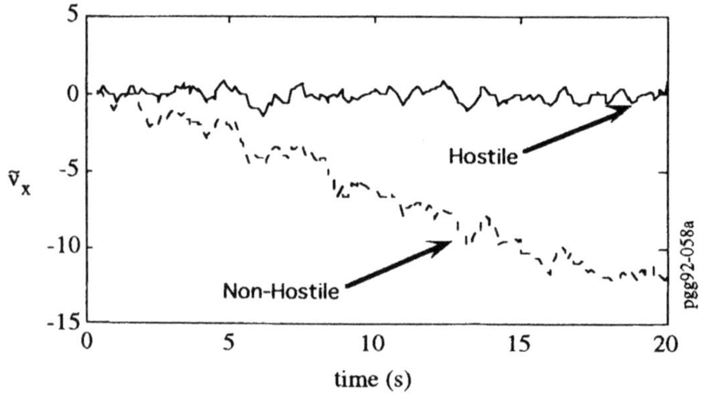

Figure 8.11 Residual-Based ANN Features - x-Velocity

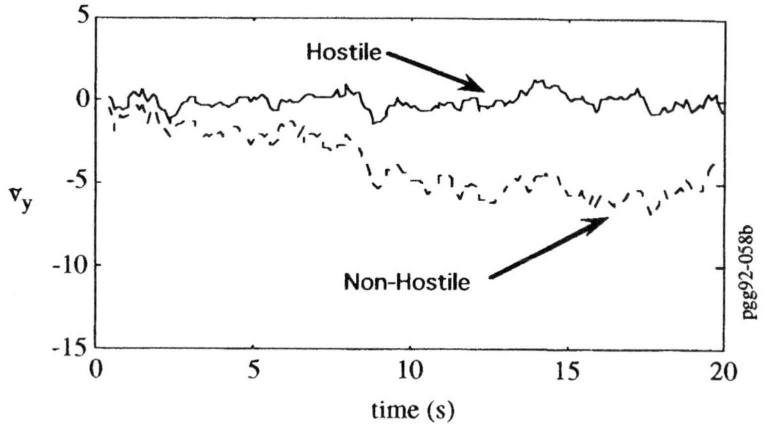

Figure 8.12 Residual-Based ANN Features - y-Velocity

measurement residual at frame k. These measurement residuals scaled by the standard deviation of the measurement noise, and averaged over a moving window are used as feature inputs into an ANN.

Figure 8.11 shows the x-velocity residuals-based ANN features for both a hostile and a non-hostile target, whereas figure 8.12 shows the y-velocity ANN features. The target estimator produces a zero-mean unit-variance random sequence for a hostile target, whereas it produces a non-zero mean random sequence for a non-hostile target. This is because the estimator assumes that the target implements a turn-to-heading intercept guidance law in the computation of the target acceleration that serves as inputs into the estimator via (8.13).

The second set of features is derived directly from the sensor measurements without filtering using invariance considerations. These ANN input features are defined by:

$$\cos(\mathrm{TAA}) = \frac{\mathbf{v}}{|\mathbf{v}|} \cdot \frac{\mathbf{r}}{|\mathbf{r}|}$$

$$\tau = \frac{\dot{\rho}}{\rho} \qquad (8.15)$$

where TAA is the target aspect angle (i.e., the angle between the LOS vector and the target velocity), **v** is the target velocity vector, $|\mathbf{v}|$ is the magnitude of the target velocity vector, **r** is the target position vector, $|\mathbf{r}|$ is the magnitude of the target position vector, and "." denotes the vector dot product. The cosine of the target aspect angle is given by the dot product of the position and velocity unit vectors, τ is the inverse of the target closure rate, $\dot{\rho}$ is the target range rate, and ρ is the target range. The closure rate is given by range (ρ) divided by the range rate $(\dot{\rho})$.

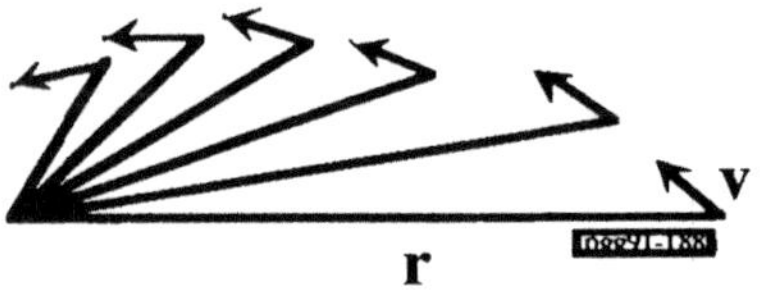

Figure 8.13 Target Maneuvers with Identical Aspect Angles

The ANN features are selected based on invariance considerations depicted in figures 8.13 and 8.14. Figure 8.13 shows a typical set of target maneuvers that would yield the same feature based on the Target Aspect Angle. This invariance property is important both in minimizing the training data and in classifying the target behavior. Figure 8.14 shows a typical set of target maneuvers that would yield the same ANN input feature based on the closure rate.

Figures 8.15 and 8.16 show both of the invariance-based features, respectively, for both a hostile and a non-hostile target. Compared to the residual-based features, the distinguishability between hostile and non-hostile targets is more immediately apparent since the residual based features require the propagation of estimator dynamics to generate features.

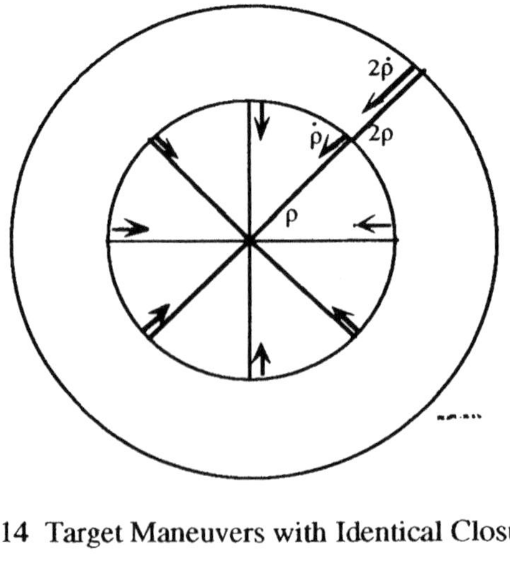

Figure 8.14 Target Maneuvers with Identical Closure Rates

The Neural Network Classifiers consists of the network using the estimator residual-based features and a network using the sensor-derived invariance-based features. For both networks a feedforward backpropagation network is used with an architecture as shown in figure 8.17. The architecture consists of processing the input features through delay operators (denoted by D where D(x(k)) = x (k-1)) to provide a moving window average of the ANN input features; a hidden layer with ten nodes; and a single output signal that is also processed through delay operators to provide a moving window average for target classification. The moving window average is based on four measurements. The moving windows are used to introduce filtering to enhance the ANN target classification performance. This architecture has been selected over a competing recurrent backpropagation network architecture due to its fast training performance. The ANN topology is fully-connected. The number of hidden nodes is determined by training performance and by the number of input/output pairs presented.

The next step in neural network classifier specification is the training of the networks. Training involves presenting the network input/output pairs and updating the network parameters (weights and biases) via the backpropagation learning algorithm until an acceptable level of error between desired and network outputs is achieved. At that point, the network parameters are frozen and the network evaluated for its on-line classification performance. Training data for both networks were generated for both hostile and non-hostile targets. Both networks were trained such that the output node was activated for a hostile trajectory and inhibited for a friendly trajectory.

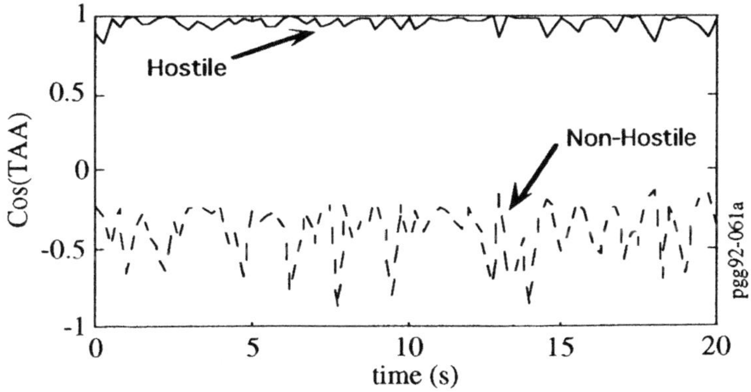

Figure 8.15 Invariance-Based ANN Input Features - Cosine of TAA

The residual-Based ANN was initially trained with input data similar to that shown in figures 8.13 and 8.14. However, this training set was not sufficient for acceptable generalization. For instance, a $\tilde{V}_x$ and $\tilde{V}_y$ residual input pair of -20,0 (in units of standard deviations) produced a friendly result, but the input pair 0,-20 did not. This example clearly shows the lack of invariance in the selected cartesian coordinate system based variables.

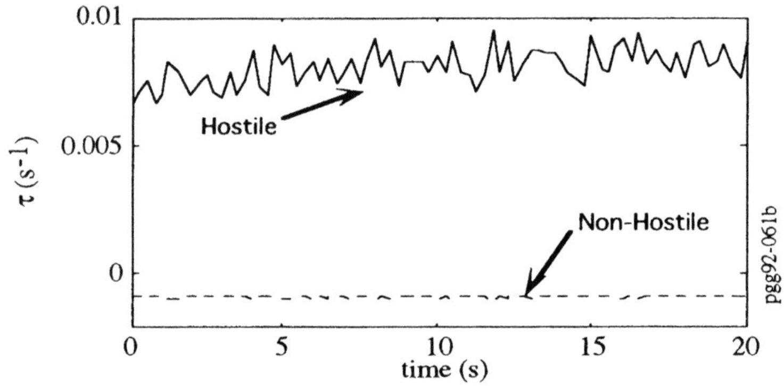

Figure 8.16 Invariance-Based ANN Input Features - Inverse Closure Rate

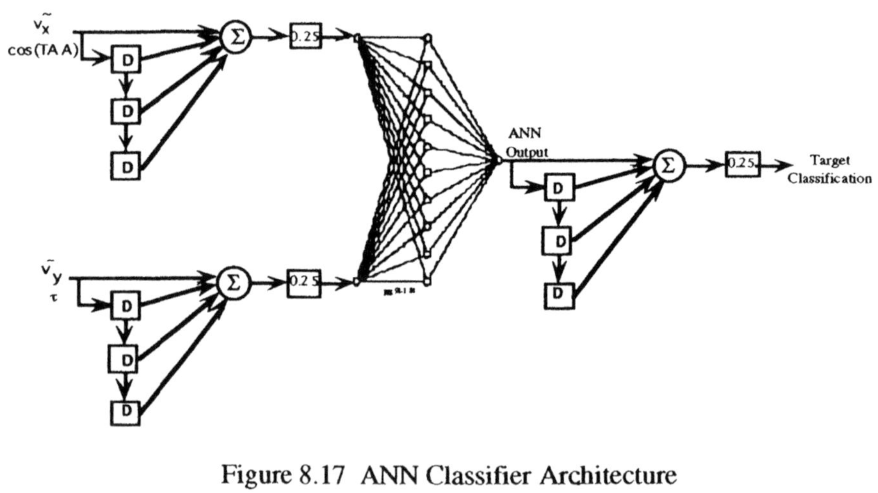

Figure 8.17 ANN Classifier Architecture

Based on these results, we took a different approach to the residual-based network training. Instead of using trajectory data, we used randomly, uniformly distributed data points in the residual $\tilde{V}_x$, $\tilde{V}_y$ space for hostile and friendly targets, respectively. Hostiles correspond to points within the region given by both residuals having absolute values less than 2.5 standard deviations whereas that region of points having at least one of the two residuals with absolute value greater than 4.0 standard deviations correspond to friendlies. Clearly, we could have used a Bayesian classifier with similar performance. The region between 2.5 and 4.0 standard deviations was omitted so as to not present conflicting data to the network. Symmetry was also ensured by taking the initial randomly selected points and creating mirror images in the three other quadrants (i.e., for a pair $(\tilde{V}_x, \tilde{V}_y)$, $(-\tilde{V}_x, -\tilde{V}_y)$, $(-\tilde{V}_x, \tilde{V}_y)$, and $(\tilde{V}_x, -\tilde{V}_y)$ were also generated). The resulting input/output pairs were scaled between 0.1 and 0.9 and then used to train the network.

The network was trained for 25,000 cycles with the resulting maximum error of 0.1234 and a root-mean-squared (RMS) error of 0.038 for the scaled training data. Network training was stopped after 25,000 cycles since the RMS training error reached a steady-state value, and the maximum training error was less than the desired upper limit of 0.2. We have selected 0.2 as the desired maximum value since we classify any network output greater than 0.7 as an hostile, and any network output less than 0.3 as friendly.

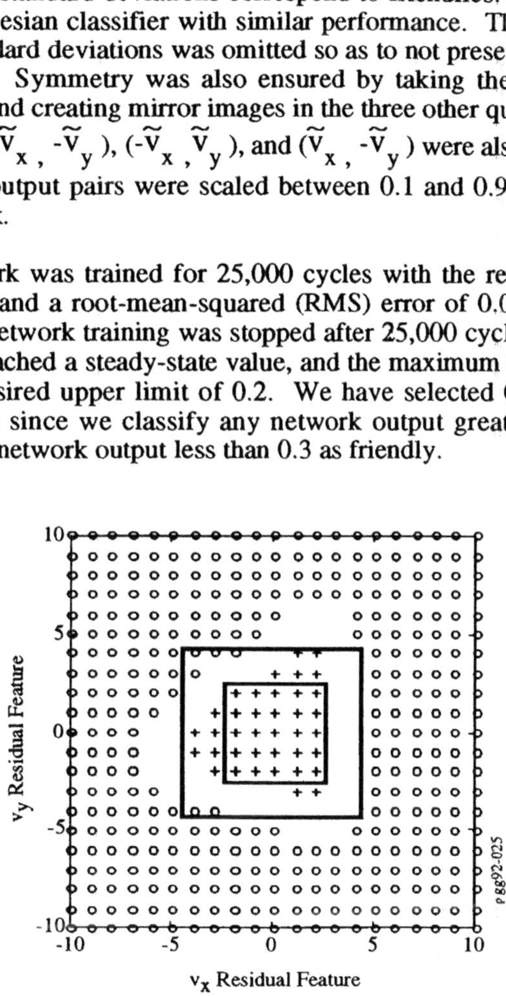

Figure 8.18 Residual Network Decision Space

The network with its associated parameters was then analyzed by investigating the input/output relationships of the network, or equivalently looking at the decision space of the network. Figure 8.18 shows the decision space for the trained network. The $\tilde{V}_x$, $\tilde{V}_y$ region highlighted with a "+" symbol correspond to the inputs that will activate the output node, or equivalently a hostile classification. The region denoted with the "o" symbol corresponds to the input that will inhibit the output node, or equivalently a friendly classification. The remaining points in the decision space correspond to a friendly classification. As shown, a hostile classification results when both $\tilde{V}_x$ and $\tilde{V}_y$ are within the +/- 3 standard deviations region with a few point points stretching to 4 standard deviations. A friendly classification occurs when one or both of the input features exceed 5 or 6 standard deviations.

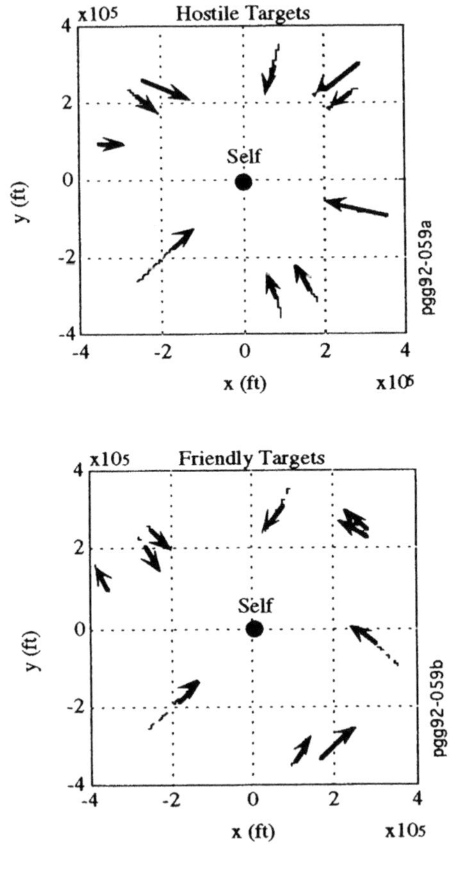

Figure 8.19 Target Scenario for Invariance-Based Feature Evaluation

The invariance-based neural network was trained using trajectory data. The trajectories employed in training and evaluation employ twenty objects with ten objects heading directly toward self (hostile targets) and the other ten heading in various other directions. Figure 8.19 shows the geometry for the scenarios considered. For training purposes, two hostile trajectories and two friendly target trajectories were used.

The network was trained using input/output pairs from training input data for 25,000 cycles. The resulting maximum and RMS errors for the scaled data were 0.829 and 0.0048, respectively. This network was able to train to lower error levels than the residual-based network because the region in the decision space where hostiles and friendlies are determined is more easily defined due to invariance. This is verified by inspecting figure 8.20 which shows the decision hyperplanes in the feature space. The points denoted by "*" correspond to an unknown classification. The region within these points results in a hostile classification. The region outside correspond to a friendly target classification.

This example clearly shows that the neural networks can replace the binary hypothesis, sequential probability ratio, multiple hypothesis tests in MTR functions. Furthermore, the ANN based approach is basically a thresholdless approach where the decision hyperplanes (lines in this example) in the feature space are automatically generated by the network learning process.

Classification Knowledge Base: The classification knowledge base provides the functionality to encode domain specific expert reasoning to the problem of target classification. The knowledge base incorporates this domain specific expertise into rules which are composed in an IF-THEN structure. The rules are implemented to perform target classification using sensor derived target data input features.

The specific rules used in the knowledge base classifier are documented in table 8.1. The table lists the IF and THEN parts of the rules in order of firing sequence priority, i.e., in the order which rules would be fired first. The rule set consists of nine rules altogether. The first five rules declare the target friendly or hostile based on target range, range rate, and the cosine of the Target Aspect Angle (TAA). The cos(TAA) provides an indication of the target's heading, whereas range and range rate provide indicators of proximity and closure rate. The specific numbers used in the conditional part of these rules are derived from an assumption of how fast the target can reach a danger region assumed to be within a 30 nm range. For instance, in rule 1, the current knowledge base output is set to hostile if target range is between 55 and 60 nm, its heading is generally towards self, and it can reach the danger region within at least 5 minutes. For closer targets, this time to danger region is increased because proximity becomes more of a concern. The current knowledge base output is set to friendly when targets appear to be moving away from self or acting in a non-hostile manner.

Rules 6 and 7 account for the possibility that a hostile target could be performing a maneuver when moving away from self. That is, a hostile target could be turning in order to acquire ownship. The manner in which these rules make this determination is by inspecting a knowledge base feature called maneuver_detect based on the time rate of change of the cos(TAA) or mathematically:

$$\text{maneuver\_detect} = \frac{d(\cos(TAA))}{dt} \cdot \frac{|v|}{\rho} \tag{8.16}$$

The derivative term is scaled by the target's speed and range in order to emphasize targets traveling at higher speeds and at closer ranges.

The final two rules of the classification knowledge base assert decisions passed to the executive knowledge base. If the target has been asserted hostile three of the four last times by the knowledge base as determined in rules 1 through 5, the target is classified a hostile. If it has been classified friendly three out of the last four times and it is not maneuvering, the classification knowledge base classifies the target as a friendly. If neither of these rules apply, no classification is made and the target is still considered unknown.

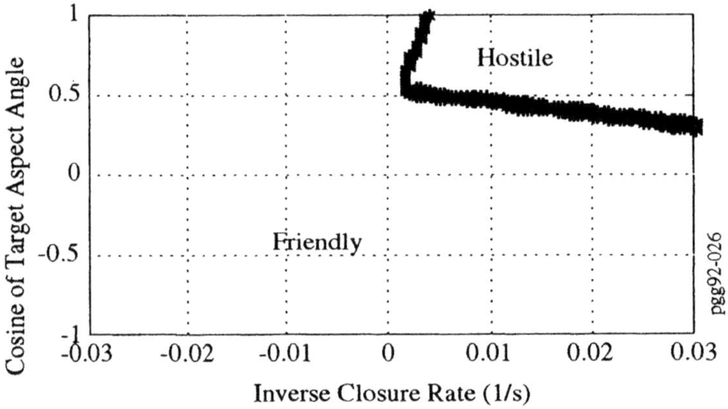

Figure 8.20 Decision Hyperplanes in Invariance Feature Space

Executive Knowledge Base: The overall target classification decision-making is performed in the executive knowledge base. This knowledge base interrogates one of two ANNs (either the residual-based features or invariant-based features network) and the classification knowledge base and uses ANN, KB, and other information to make a final decision on target status. The specific rules implemented are listed in table 8.2 in order of rule firing priority.

The three rules with the highest rule firing priority perform target classification based on information other than in the classification knowledge base or the ANNs. Specifically, these rules deal with information the ship commander may have from other sources such as Identification Friend or Foe (IFF) systems or making the determination that if a target has radar lockon on ownship, then it is hostile. The third rule accounts for targets that are initially within sensor range, but than go out of range and are never detected again. These targets are assumed friendlies due to their non-hostile trajectories (i.e., their heading directly away from ownship).

Rule 4 has the next highest priority level. It basically serves as a flag to determine if the current target has been classified and if it is within sensor range. If both conditions are true, then the classification procedure for the target will be implemented. This procedure is set forth in rules 5 through 7. Here, the executive knowledge base interrogates the classification knowledge base and the selected ANN. Rules 8 and 9 take the average of the last four outputs of the selected ANN and make a classification decision according to the ANN.

The overall MTR system target classification decision-making is implemented in rules 10 through 13. In these rules, comparisons are made between the results of the ANN and the classification knowledge base, and based on these comparisons overall decisions are made. A target is considered hostile if either the ANN or the knowledge base returns hostile while the other is declaring anything but friendly. This logic is implemented in rules 10 and 11. A classification of hostile is also made under the special circumstance when both ANN and the knowledge base are declaring unknown and the target is within the danger region specified by a range of less than 30 nm. Rule 12 implements this logic. A classification of friendly is made only when both the ANN and the knowledge base are returning friendly as specified in rule 13.

NueX Implementation of MTR System: The MTR system prototype has been implemented using our in-house hybrid development tool NueX. The overall system is implemented in NueX via a HyperCard stack. The stack integrates the functionality of NueX with the graphical capabilities of HyperCard. NueX provides the capabilities for the neural network and the knowledge bases. The MATLAB generated trajectory data files containing sensor and filter data are read and processed within the system using the HyperCard scripting language HyperTalk. HyperTalk script is also used to drive display processing.

Figure 8.21 shows a view of the MTR system display as it would appear on an Apple Macintosh screen before a run. As shown in figure 8.21, the majority of the display area is taken up by a radar display screen. The radar screen provides a two-dimensional display of position for all targets within the sensor range of 60 nm. The display consists of concentric circles at 15 nm intervals centered at a point which is at the crosshairs center (i.e., ownship current position).

Table 8.1 Knowledge Base Target Classifier

Rule Number	IF	THEN	Rule Firing* Priority
1	no target within sensor range (60 nm) & range < 55 nm & range rate < -700 ft/s & cosine(TAA) > 0.7	assert current target status is hostile	High
2	range < 55 nm & range ≥ 45 nm & range rate < -500 ft/s & cosine(TAA) > 0.8	assert current target status is hostile	High
3	range < 45nm & range ≥ 30 nm & range rate < -100 ft/s & cosine(TAA) > 0.9	assert current target status is hostile	High
4	range < 55 nm & range rate > 0 ft/s & cosine(TAA) < 0	assert current target status is friendly	High
5	range < 45 nm & range rate > 0 ft/s & cosine(TAA) < 0.5	assert current target status is friendly	High
6	-4 < maneuver_detect < 4	assert target is not manuvering	Med
7	maneuver_detect > 4 or maneuver_detect < -4	assert target is manuvering	Med
8	target has been asserted hostile three of last four times	assert target is hostile	Low
9	target has been asserted friendly three of last four times & target is not maneuvering	assert target is friendly	Low

pgg92-018

*ordering precedence for firing of rules

Besides the graphical information presented by the radar screen display, the MTR display also provides textual information. There are three textual elements or *fields* located on the display: 1) information concerning the status of all targets is provided in the "Target Status" field; 2) the current simulation time is presented in the "TIME" field; and 3) relevant events are presented in the "EVENTS" field as they occur during the simulation run. The "Target Status" field contains information on each of the targets as to what their real classification is and what the ANN, the classification knowledge base, and the executive knowledge base have classified it. Information in the "EVENTS" field consists of when and how the MTR system came to a make classification.

The final aspects of the MTR display screen include a *button* that the user can toggle to select either the residual-based (Residual Net) or the invariant-based features (Features Net) ANN and the menus located at the top of the screen. The user selects which ANN to use by clicking the mouse on the circle to the left of the desired network. The circle or *button* then becomes highlighted. The menus include all the menus available with NueX and the added "Simulation" menu. Under the "Simulation" menu, the user can either make a run by choosing "Run" or can clear the display screen and the program memory after aborting a run. Once the user selects "Run", he is then prompted for an input data file. He is then prompted for the number of targets in that data file. Finally, the user is prompted for an output file name and log file name. The simulation writes the overall MTR system target classifications to the output file. The information presented in the "EVENTS" field during a run are saved to the log file.

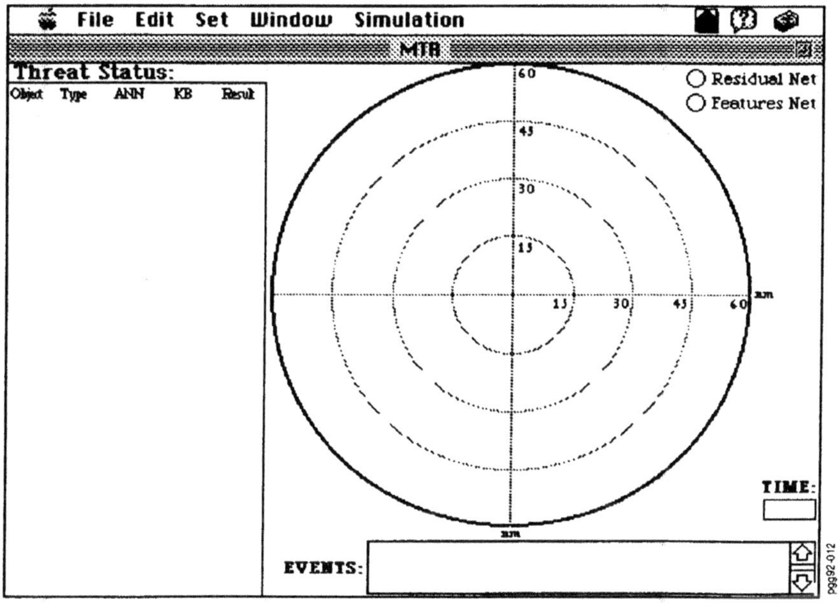

Figure 8.21 MTR Display Interface

8.4 Trade Studies

Trade studies were performed to evaluate the performance of the prototype hybrid MTR system described in section 8.3. Our goal is to demonstrate the feasibility of a such a hybrid system approach to multiple target recognition. Specific items to be considered include performance under varying radar sensor

Figure 8.2 Executive Knowledge Base

Rule Number	IF	THEN	Rule Firing* Priority
1	target has not been classified & target within sensor range (60 nm) & IFF on	assert target is friendly due to IFF	Very High
2	target has not been classified & target within sensor range (60 nm) & target has achieved radar lock-on	assert target is hostile due to radar lock-on	Very High
3	target has not been classified & simulation time > 10 s & target within sensor range less than the half of the time	assert target is friendly due to going out of sensor range	Very High
4	target has not been classified & target within sensor range (60 nm) & cosine(TAA) < 0	assert classify target	High
5	classifying target & residual-based ANN selected	assert cycle residual-feature ANN	Medium
6	classifying target & invariance-based ANN selected	assert cycle invariance-based ANN	Medium
7	classifying target	assert run classification knowledge base	Medium
8	classifying target & target has been asserted hostile three out of the last four times by ANN	assert ANN classification of target is hostile	Low
9	classifying target & target has been asserted friendly three out of the last times by ANN	assert ANN classification of target is friendly	Low
10	classifying target & ANN classification is hostile & knowledge base classification is not friendly	assert target is hostile	Very Low
11	classifying target & knowledge base classification is hostile & ANN classification is not friendly	assert target is hostile	Very Low
12	classifying target & knowledge base classification is unknown & ANN classification is unknown & range < 30 nm	assert target is hostile	Very Low
13	classifying target & knowledge base classification is friendly & ANN classification is friendly	assert target is friendly	Very Low

*ordering precedence for firing rules

pg92-019

signal-to-noise ratios (SNR) and varying target density. For each variation, the performance of the following systems are compared:

- Performance of hybrid architecture using residuals ANN vs. that using invariance ANN
- Performance of ANN classifier vs. KB classifier vs. Hybrid Classifier

Performance is judged based on three metrics: 1) probability of missed detection; 2) probability of false alarm; and 3) average detection time for hostile targets. For each variation, object trajectories have been generated and sensor data and input features gathered using our MATLAB-based simulation facility. A nominal set of parameters were used to generate these trajectories and are listed in table 8.3. Two assumptions made in these trajectories were that ownship is stationary and target trajectories lie in the same plane as ownship, i.e., we constrain ourselves to the two-dimensional MTR problem. The trajectory initial conditions for both friendlies and hostile targets were randomly selected over a typical range commensurate with capabilities of existing US carrier-based aircraft and for surface attack hostile aircraft and missiles.

Table 8.3 Nominal Parameter Set

Parameter	**Value**
Object Density	25
Number of Hostiles	15
Number of Friendlies	10
Radar Sensor Range	60 nm
Radar Sensor SNR	25 db
Azimuth/Azimuth Rate Noise Level	1°, .1°/s

Table 8.4 Trade Studies Performance Metrics

METRIC	CALCULATION
Probability of Missed Detection (P_{md})	$\frac{\text{No. of hostiles classified friendly or unknown}}{\text{No. of hostile threats}}$
Probability of False Alarm (P_{fa})	$\frac{\text{No. of friendlies classified hostile}}{\text{No. of friendlies}}$
Average Detection Time (T_d)	$T_d = \frac{1}{\text{No. of hostiles classified correctly}} \sum_i T_{d_i}$

The three metrics used to evaluate performance of the hybrid MTR system as stated previously are the probabilities of missed detection and false alarm, and the average detection time for hostiles. For this study, we concern ourselves only with *initial* target classification, that is we calculate statistics concerning only the system's first target classification. Thus, all results presented pertain to how well the system performs in initially classifying targets. Once the system has classified a target, that target is no longer a candidate for classification. The performance metric are thus defined as shown in table 8.4. As shown, a missed detection occurs if a hostile target is initially classified friendly or no classifca-tion is made (i.e., unknown classification). A false alarm is generated when a friendly is classified as a hostile. The average detection time for hostiles (T_d) is the average time of detection over all the hostiles classified correctly. This statistic provides an indication of how quickly the system performs.

8.4.1 Impact of Radar Signal-to-Noise Ratio Variation

In this section we investigate the MTR system performance under varying radar sensor noise levels. Results are presented for a range of radar sensor signal-to-ratios starting from low noise levels (40 db) to high noise levels (10 db). The results for each SNR value are averages over three simulation runs generated with trajectory files having different initial conditions. To offset initial transients in sensor and filter data, delays were introduced into the simulation runs. For the invariance-based features, the first two time ticks were ignored. For the residual-based features, the first 21 time ticks (or the first five seconds of simulation time using a simulation time step of 0.25 s) were ignored to allow the residuals to ramp up and thus to improve false alarm performance. Figures 8.22 and 8.23 present the missed detection and false alarm performance, respectively, for the hybrid system using the invariance-based ANN. Comparable results using the residual-based ANN are shown in figures 8.24 and 8.25. Included in these figures is not only the overall hybrid system performance (as denoted by MTR), but performance of the neural network (denoted by ANN) and the classification knowledge base (denoted by KB) are also presented for comparison.

As shown in figures 8.22 and 8.23, the hybrid system using the invariance-based network performs well with missed detections usually well under 10% and false alarms around 10% over the range of radar sensor SNR. The only exception is the highest noise case (10 db) where false alarms are 40%. The missed detection performance of hybrid system consistently surpasses that of the individual ANN and KB classifiers. False alarm performance of MTR will always be only as good as the worst of the individual classifiers based on the logic implemented in the executive knowledge base as detailed in table 8.2. The reason for the very high system false alarm rate for the 10 db case is due to high range noise making targets appear much closer and thus the executive knowledge base asserts a hostile classification as per rule 12 in table 8.2.

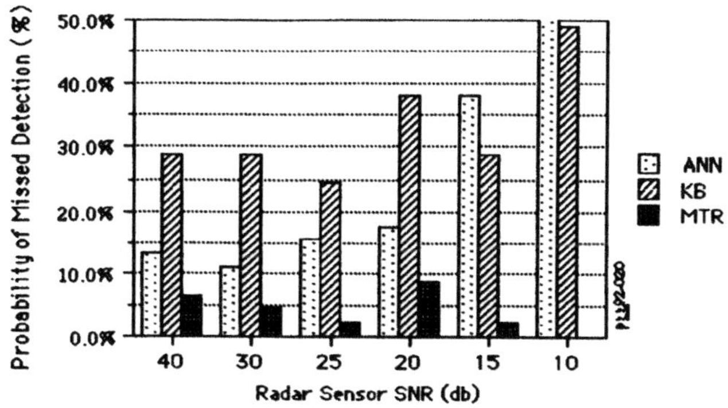

Figure 8.22 Effect of SNR Variation on Missed Detection Performance *Invariance*-Based Features

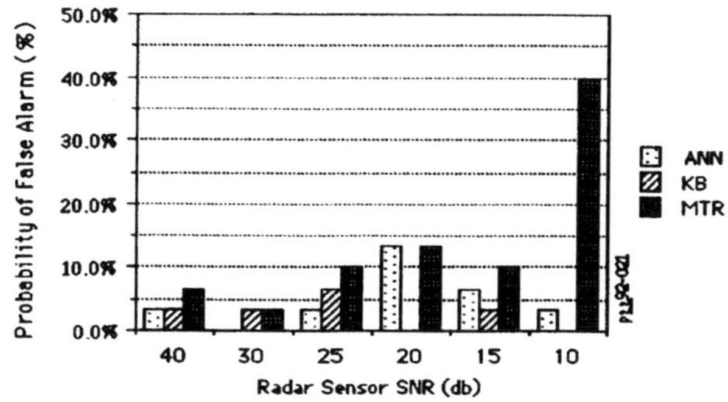

Figure 8.23 Effect of SNR Variation on False Alarm Performance *Invariance*-Based Features

The MTR system using the residual-based features achieved very good hostile target classification performance as shown by the missed detection probabilities in figure 8.24. However, the false alarms for the hybrid system are high with values around 20% and as high as 40% as shown in figure 8.25. The classification knowledge base is seen to perform comparably with both sets of ANN features. The cause for these false alarms is due to the performance of the residual-based feature network. The problem lies not with ANN itself, but with

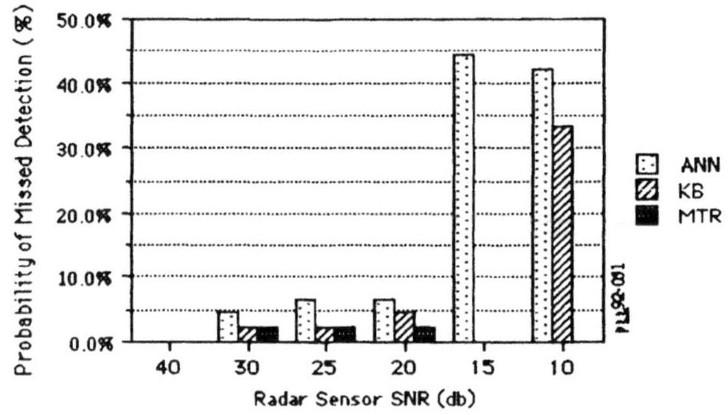

Figure 8.24 Effect of SNR Variation on Missed Detection Performance *Residual*-Based Features

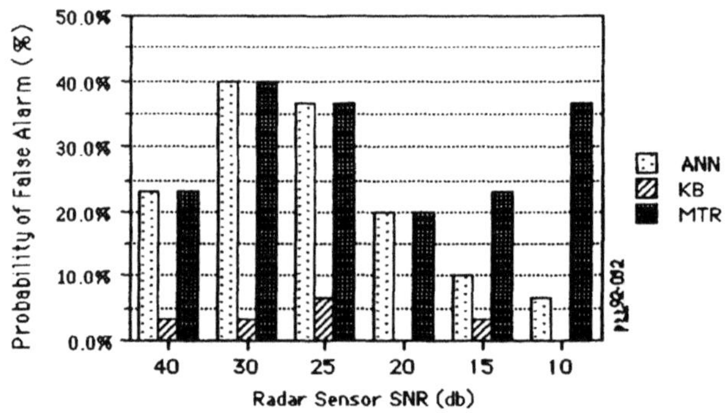

Figure 8.25 Effect of SNR Variation on False Alarm Performance *Residual*-Based Features

the hypothesis being made in the implementation. That is, the hypothesis that all targets are using the turn-to-heading intercept. This demonstrates the disadvantages of model-based hypothesis testing. Modeling discrepancies can lead to degraded system performance, whereas the invariance-based features make no modeling assumptions and thus are more robust.

The hybrid system average hostile detection time performance for both sets of features for varying SNR is shown in figure 8.26. The invariance-based system outperforms the residual-based across all the variation in SNR. The difference usually being by a factor of three. The main reason for the enhanced performance for the invariance-based system is the delay required by the residual-based system to allow the residuals to ramp up.

8.4.2 Target Density Variation

The effect of an increased target environment or target density on the hybrid system classification performance is now discussed. Target density can directly impinge on system performance. Modeling of increased target density can be accomplished by increasing sensor noise. This, however has been investigated in the previous section. We implement an increased target density level by making the assumption that the MTR system can only process a limited number of targets per computation frame. This assumption is useful to account for real-time sensor and processing capabilities as well as sensor drop offs. Results are presented for target densities of 25 (nominal case), 50, 75, and 100. All other parameters are as in the nominal case. Results are generated for both the residual-based and invariance-based features.

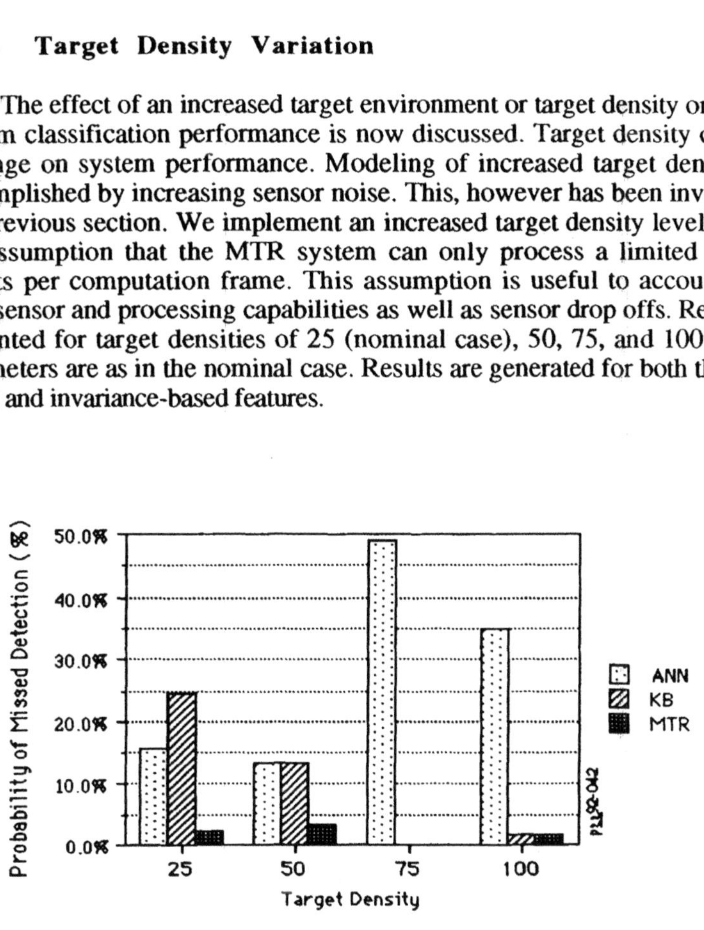

Figure 8.26 Effect of Target Density Variation on Missed Detection Performance-*Invariance*-Based Features

Figures 8.27 through 8.31 detail the simulation performance results for invariance-based and the residual-based features, respectively. As shown in figure 8.2, the missed detection performance of the invariance-based features system remains consistent across target density. However, the ANN missed detection performance deteriorates at the higher density levels, in part due to the ANN using initial data during which hostiles are still maneuvering (i.e. , they may be turning about to face ownship). This indicates that the addition of other features such as the maneuver detector used in the KB classifier can improve the performance of the ANN classifier

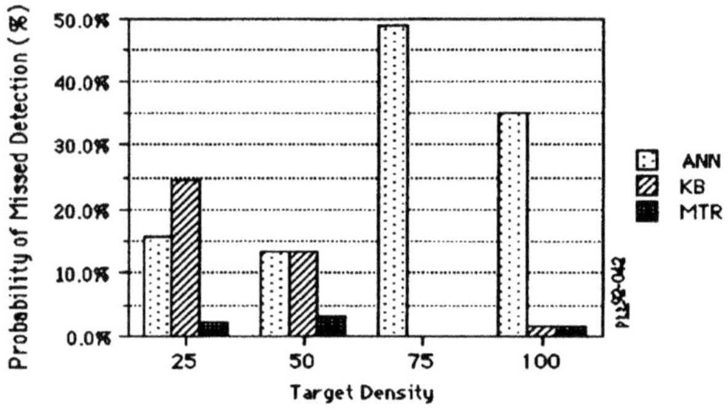

Figure 8.27 Effect of Target Density Variation on Missed Detection Performance-*Invariance*-Based Features

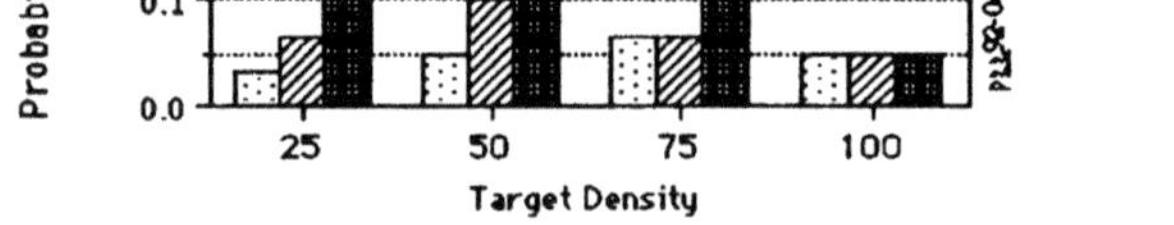

Figure 8.28 Effect of Target Density Variation on False Alarm Performance -*Invariance*-Based Features

MTR system false alarm performance is also generally consistent across the variation as shown in figure 8.28. For the residual-based features, no trends are noticeable for missed detection as shown in figure 8.29. False performance is seen to improve at the higher density levels as shown in figure 8.30 with false alarm probability dropping about 15% from the 50 target case to the 75 and 100

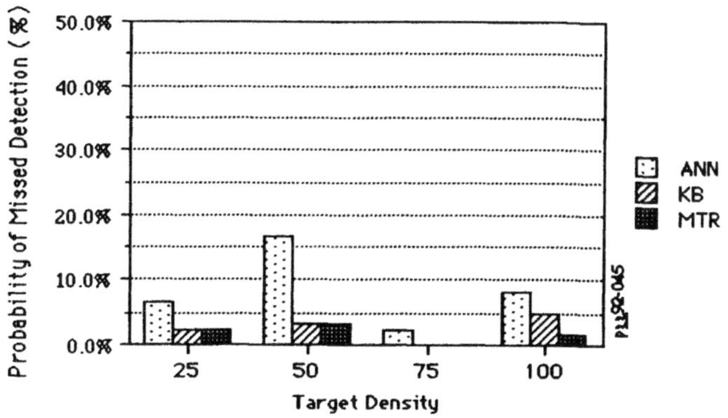

Figure 8.29 Effect of Target Density Variation on Missed Detection Performance-*Residual*-Based Features

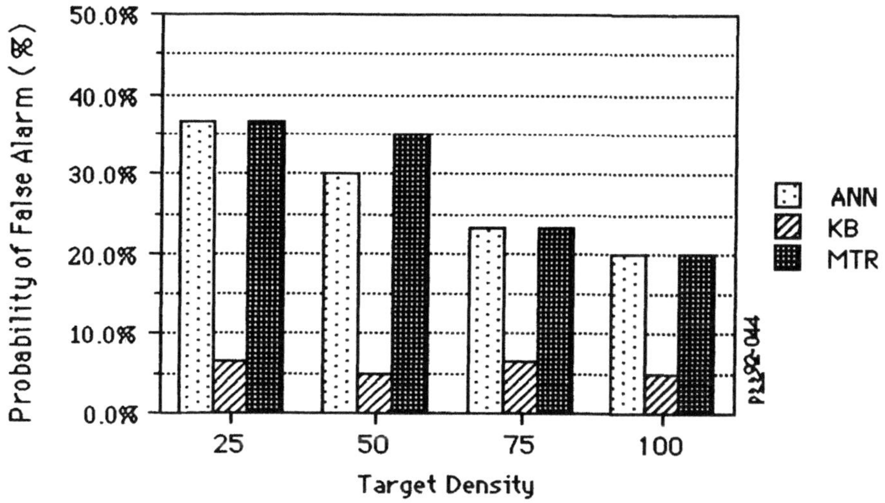

Figure 8.30 Effect of Target Density Variation on False Alarm Performance -*Residual*-Based Features

target cases. This decrease is due in part to the added delay allowing the residuals to ramp up and thus decrease the number of friendlies classified as hostiles by the ANN. Finally, the average detection times for hostile classifications across the target density variation for both sets of features are shown in figure 8.31. Again, the invariance-based outperforms the residual-based features across the variation by over a factor of three at the lower density levels to a factor of 2 at the high density levels. Both sets of features demonstrate increasing detection times with increasing target density due to the added delays.

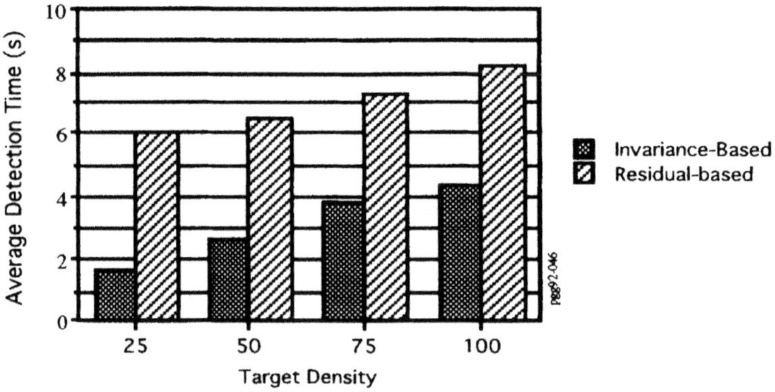

Figure 8.31 Effect of Target Density Variation on Average Detection Time

8.5 Summary, Conclusions & Recommendations

Future work should include the development of a full-scale research prototype for a hybrid MTR system expanding the prototype MTR architecture. Such an effort would address the following:

- Incorporation of three dimensional (3D) scenarios into the architecture: The Phase I study has been limited to 2D scenarios. The inclusion of 3D trajectories would necessitate the development of additional features based on invariance. Similarly, additional preprocessing would be required for the knowledge base classifier inputs.

- Expansion of the sensor suite: The prototype has been limited to the processing of radar data. Under a Phase II development, the architecture would be modified to focus on sensor capability for over-the-horizon (OTH) radar with a more detailed measurement system simulation. This would necessitate the consideration of other neural network paradigms such as Kohonen or ART networks more suited to the image processing problems. The inclusion of data with uncertainty would require the inclusion of reasoning algorithms such as Shafer-Dempster (Shafer and Logan, 1987), fuzzy logic (Zadeh, 1986), or Bayesian networks (Charniak, 1991) more suitable to reasoning under uncertainty.

- Expansion of the hybrid architecture to include MTR systems composed of neural networks based on different paradigms and knowledge based expert systems with different inference protocols: This would be desirable not only to handle additional sensor data but also to improve the performance of the MTR system.

- Determination of the hybrid MTR system performance using a more realistic simulation environment: This would entail measuring the functional performance of the full-scale research prototype system using differing external environments and with varying sensor suite capabilities.

- Determination of real-time operation requirements: This would entail finding the real-time operational requirements of the MTR system relating to I/O requirements, memory allocation, throughput, duty cycles and other factors.

Part III Analysis and Guidelines

Chapter 9

Guidelines for Developing Hybrid Systems

9.1 Characteristics of Expert System and Neural Network

Expert systems perform reasoning using pre-established rules for a well-defined and narrow domain. They combine knowledge bases of rules and domain-specific facts with information from clients or users about specific instances of problems in the knowledge domains of the expert systems. Ideally, reasoning can be explained and the knowledge bases easily modified, independent of the inference engine, as new rules become know.

Expert systems are especially good for closed-system applications for which inputs are literal and precise, leading to logical outputs. They are especially useful for interacting with the client/user to define a specific problem and bring in facts peculiar to the problem being solved. ES reasons using established facts and pre-established rules.

A major limitation of the expert system approach arises from the fact that experts do not always think in terms of rules. Thus, an expert system does not in such cases mimic the actual reasoning process of human experts. For example, credit application evaluation involves aspects such as human factors and organizational issues; theses are difficult to be captured in rules, but appear as text in the application processing approval.

Neural networks rely on training data to "program" the systems. Particular applications are developed by establishing an appropriate training set that allows the system to learn and generalize for operation on future input data. Inputs that exactly match the training data are recognized and identified, while new data (or incomplete and noisy versions of the training data) can be put into the closest matches to patterns recognized by the system.

For stable applications with well-defined rules, expert systems can be easily developed to provide good performance. Furthermore, most development software packages allow the creation of explanations to help the user understand questions being asked or conclusions and reasoning processes. In contrast, in ANN, knowledge is represented as numeric weights, and therefore, the rules and the reasoning process are not readily explainable.

Neural networks can be preferable to expert systems when rules are not known, either because the topic is too complex or no human expert is available. If training data can be generated, the system may be able to learn enough information to function as well as, or better than, an expert system. This approach also has the benefit of easy maintenance. That is, modifications are exercised by retraining with an updated data set, thus eliminating programming changes and rule reconstruction. The data-driven property of neural networks allows adjustment of changing environments and events. Expert systems are knowledge driven so changes can be made only if the knowledge is being changed. Another advantage of the neural network implementation is the speed of operation after the network is trained; the natural use of parallel systems and neural chips enhances this aspect dramatically.

The differences between expert systems and neural networks require care in choosing the appropriate technology for a particular application. The two technologies may represent complementary or competing approaches. ANN can be the best solution for some of the problems that have proven difficult for expert system developers. The special characteristics of the technologies create opportunities for integrated systems that allow system developers to address problems not amendable to either approach alone.

Several areas involving expert systems and neural networks are thus interesting:

- Expert system elements implemented with neural networks; i.e., use of neural networks in situations where expert systems have previously been used.
- Traditional areas of AI such as speech and visual recognition tasks that can be done as well or better with neural networks
- Integrated intelligent systems -- hybrid systems with expert system and neural network components

- Novel applications that become possible because of the combination of technologies
- Interface issues --- improved user-machine interactions, as well as the use of expert systems to train neural networks
- Knowledge acquisition and knowledge engineering problems that can be addressed using neural network technology; e.g., identification of implicit knowledge by neural networks to supplement explicit rule-based knowledge.

A generic problem (see Figure 9.1) is the support of decision making in an organization by analyzing a corporate database. As shown, the data is pre-processed and used in two ways. First, when training the neural network, human intervention is required to select the appropriate training data subset and to monitor the training to ensure valid performance. Intervention is also required when retraining becomes necessary. The second use of the data is in the routine operation using the trained neural network and the expert system as a hybrid system. Again, human intervention is needed to specify problem details and requirements. The expert system component calls upon the neural network to select employees that fall into desired categories. For the selected employees, additional information is analyzed in terms of the rules in the knowledge base so that a recommendation can be made. The expert system also serves as the intelligent interface to the user and as a means of explaining the recommendations.

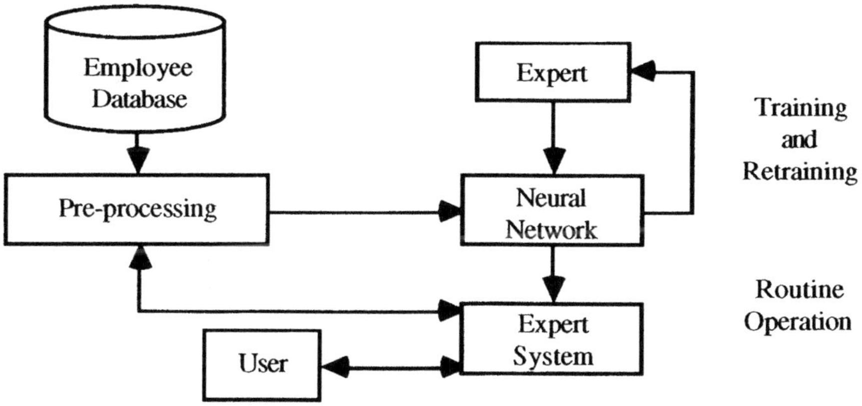

Figure 9.1. Analysis of an employee database with a hybrid system

This hybrid system approach provides several advantages that help justify this design decision (see Bailey, Feinstein, and Thompson, 1988; Bailey and Thompson, 1990; and Caudill, 1990). Among the features of this type of system are

- efficient processing of large databases,
- detailed queries or rules are not required; historical data and empirical observations are adequate,
- the system is readily updated via retraining with fresh data,
- the system can extrapolate beyond the explicit training data to produce correct results for cases not previously seen.

9.2 Hybrid System Development Issues and Methodology

As researchers and practioners are gaining experience with hybrid neural network and expert systems, models and guidelines are starting to emerge. This section describes one view of the issues and tips involved in systems development. As part of Intelligent Information Systems, the hybrid system considerations should be taken in the context of an overall system development methodology where a hybrid component is another important option that can be included in the system design.

9.2.1 Concept and Design Issues

Integrating expert systems with neural networks is a conceptually rich process, full of possibilities and decisions. More fully integrated models require a significant amount of consideration during the conceptualization and design phases of development. Two primary considerations are how to represent the knowledge within the model, and how to conduct the inferencing. Representing integrated expert systems and neural networks is more complex than representing either technology alone. For more information, please see Medsker and Bailey, 1992.

Three basic approaches are available for representing an integrated expert system/neural network. The most common and straightforward is to represent the individual software components independently and to communicate via a data interface. This allows the developer to build the expert system and the neural network modules independently. The only additional requirement is to map the data needs from one system to the other.

The second type of integration involves linking a component of the expert system knowledge representation -- such as a fact, a frame, or an object -- with a node in the neural network. This allows both the expert system and the neural network to access the same information simultaneously. When either the neural network or the expert system requires a value, it simply updates the symbol/node and proceeds.

A third form of representation bypasses the linking process entirely. Instead, the expert system symbol is the neural network node. Thus, there is no need to link the two representations, because they are one and the same. This has the advantage of clarity and efficiency in terms of computer resources, but requires more work in terms of the inferencing process. The update mechanism is responsible for keeping the value of a symbol or node current. If either the expert system or the neural network modifies a value, the update mechanism must keep the knowledge representation consistent and ensure that the other components are aware of the change.

Inferencing within an integrated system is either controlled by one of the components or occurs in a distributed fashion. Most systems designate an expert system to switch between modules based on their performance, or a neural network to focus attention and guide search. Cooperative inferencing is based on loose interaction guidelines, resource allocations, or workplans.

Adding neural networks to expert system forward or backward chaining enhances the inferencing process by adding another option for the system to explore. For example, if in the process of backward chaining the expert system's inference engine is unable to determine the value for a fact using rules or working memory, it might present the current facts to a neural network and receive an approximate classification.

Other issues arise in inferencing when developing an integrated expert. For example, when should the neural network be trained? Depending on the situation, training can occur prior to operation, incrementally, or periodically. Another consideration is how to combine the activation levels of a neural network with the uncertainty handling mechanisms of an expert system. Finally, the role and form of the explanation facility needs to be determined.

9.2.2 Hybrid System Architectures

A variety of control mechanisms are possible combining neural network and expert system components. The following examples, shown in Figures 9.2 and 9.3, illustrate some of the combinations; however, others are possible, and some complex applications have been developed recently using many neural network and expert system modules in the same system.

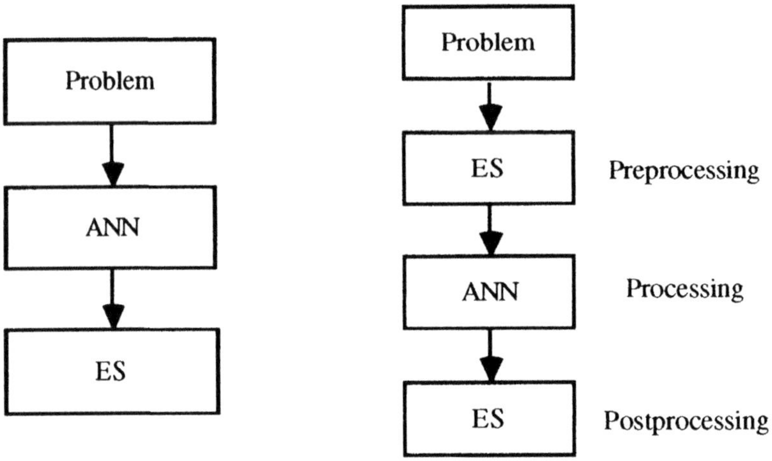

Figure 9.2. Examples of sequential configurations of neural network and expert systems.

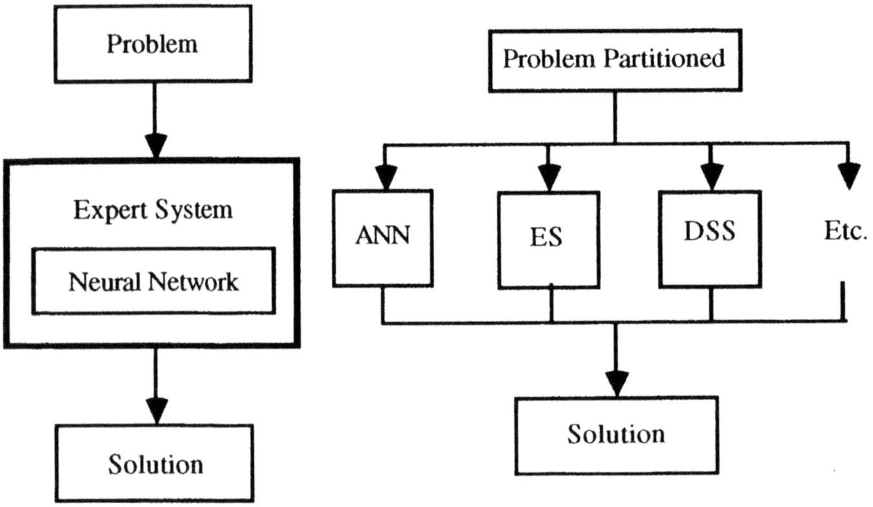

Figure 9.3. Examples of neural network and expert system configurations

In the simplest arrangement, the output of the neural network (or expert system) is input to the expert system (or neural network) in a sequential control mechanism. An additional expert system component may be used for collecting input to the neural network and/or analyzing the results.

In another configuration (see Figure 9.3), the neural network(s) are embedded in the expert system as functions to be called when needed, for example to analyze sensor data or look for trends in business data. The information gleaned by the neural network is then included with facts and rules in the expert system's reasoning process.

Of course, expert systems and neural networks can be combined with other components such as DSS's or other intelligent technologies such as genetic algorithms and case-based reasoning to make more intelligent information systems.

9.2.3 Development Strategy

Thousands of neural network tools are in use today for research and commercial applications. Effective development of large systems requires a methodology that will lead to efficient development of correct systems. The development process of any advanced computing application contains inherent risk. One means of reducing this risk is to adopt a development methodology.

The processes for developing expert systems and neural networks have important parallels as shown in Figure 9.4. An inspection of the two processes for their similarities and differences is instructive for understanding the neural network development process.

Knowledge engineers and neural network builders each have to acquire and represent a body of knowledge as facts and rules or as data sets, respectively. For supervised neural networks, data sets must be chosen accurately to represent examples of input data and their corresponding results. With unsupervised networks, a domain expert must examine and interpret the categories and perhaps recommend running the network again to obtain a different number of categories. For expert systems or neural networks, domain experts need to be involved to ensure accuracy of the rules and data sets.

The knowledge engineer spends a considerable amount of time and effort formulating and testing rules. Neural network builders have to put data sets into the proper formats for presentation to the neural network for training. This is not only a data processing problem, but also the particular format may affect the quality and efficiency of the training task.

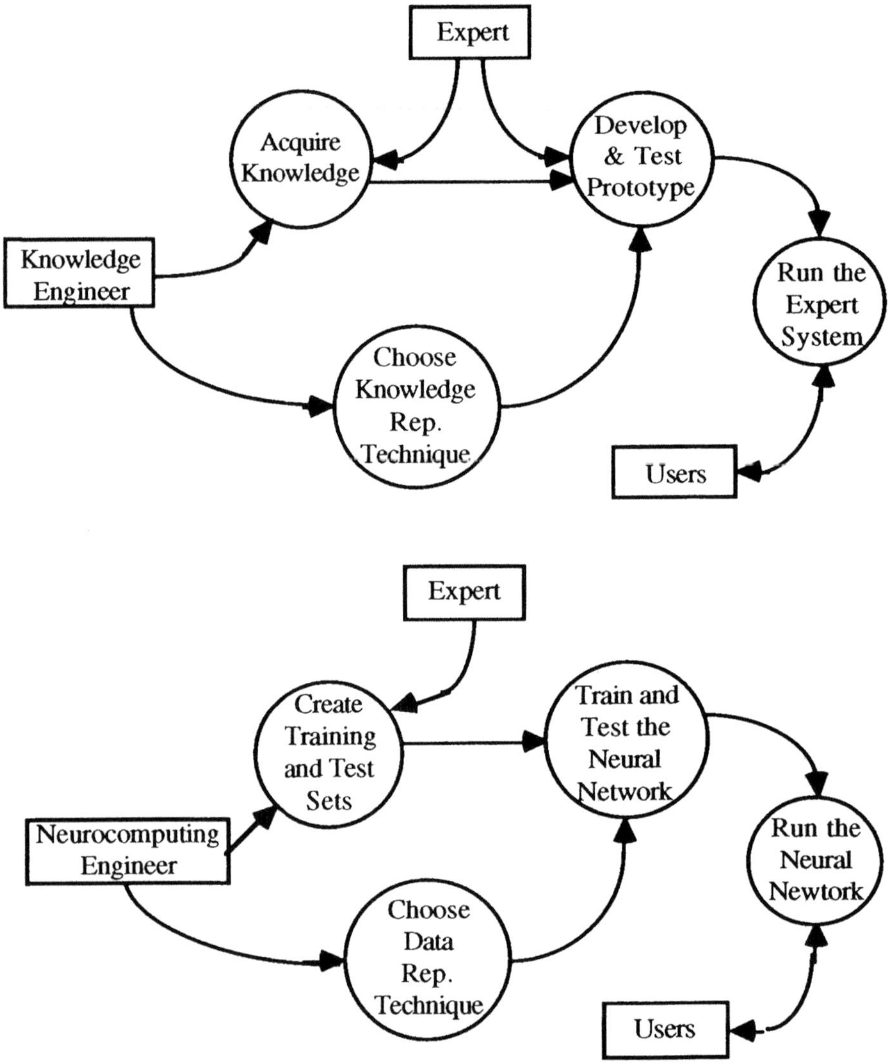

Figure 9.4 Knowledge engineering and neural computing processes

Knowledge engineers next develop the knowledge base and test the design, usually via prototyping with feedback from the expert. Neural network builders train the network to an acceptable accuracy and then run it on test cases with known results. In each case, complete validation and verification of the system is an important, but very difficult task.

9.3 Development Process

The development process of any advanced computing application contains inherent risk, and one means of reducing this risk is to adopt a development methodology. A straightforward methodology (Medsker and Bailey, 1992) for integrating neural networks and expert systems consists of four phases (see Figure 9.5): Concept, Design, Implementation, and Maintenance.

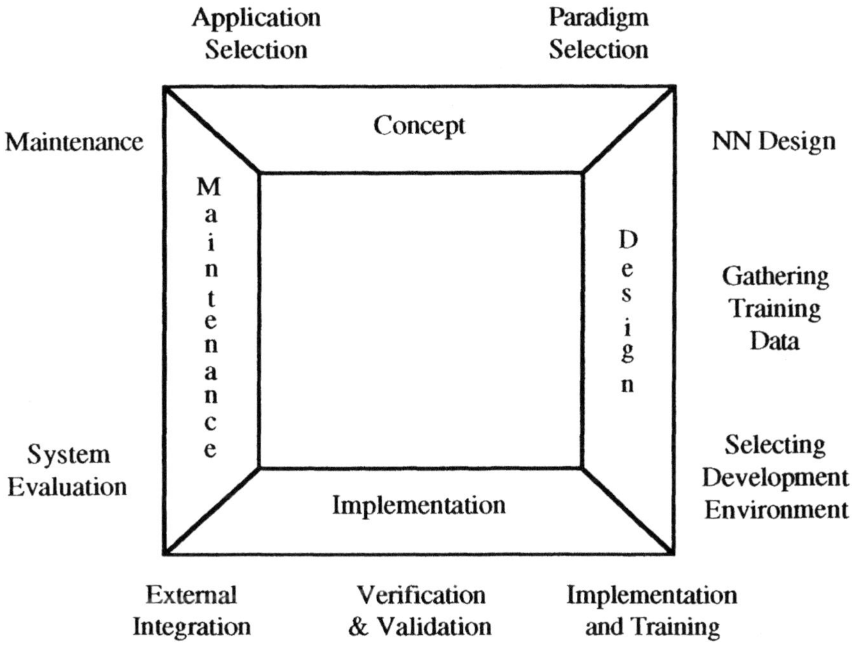

Figure 9.5. Hybrid system development process.

Within the Concept Phase, the application is selected, the problem is bounded and scoped, and the basic functionality of the system is determined. The requirements for an integrated expert system/neural network should include distinct tasks, and the level and points of required integration.

The Design Phase builds on the requirements formulated in the Concept Phase, and begins to create general and detailed designs of the individual systems as well as the integration utilities. In addition, the development environment is identified, and preliminary knowledge acquisition and data gathering is performed.

Coding the expert system and neural network environments, training the neural networks, and building the knowledge base are part of the Implementation Phase. Unit testing is performed and the units are integrated and tested. Verification and validation procedures are performed in this phase. Finally, integration with any existing, external systems is conducted. Some important activities in the implementation phase are as follows:
Data Transformation The purpose here is to transform the application data into the type and format required by the neural network. This may mean writing software for preprocessing the data. Data storage and manipulation techniques and processes need to be designed for retraining the neural network when needed conveniently and efficiently. Also, the way the application data is represented and ordered often determines the efficiency and possibly the accuracy of the neural network results.
Training This step is an iterative process of presenting input and desired output data to the network so that the weights can be determined. The previously derived test data should be used to measure to what extent the network matches the ideal performance. In order to avoid long training times, the choice of the number of nodes and layers, as well as the initial conditions of the network, is important and requires careful consideration at the outset of the process.
Unit Implementation At this point in the process, a set of weights has been obtained that allows the network to reproduce the desired outputs given inputs like those in the training set. The network is ready for use as a standalone system or as part of another software system, which could include expert system components.

The purpose of the Maintenance Phase is to evaluate periodically the system's performance, and to modify it as the need arises. Given the adaptive nature of some of the components, care must be taken to keep the entire system consistent and validated if incremental training is permitted.

Other issues are involved in developing a neural network; for example, when the neural network should be trained. Depending on the situation, training can occur prior to operation, incrementally, or periodically. If the input data are stable, the initial training produces a system that can be used from then on. An example is the routine categorizing of speech signals where all patterns of interest are known. In other applications, the characteristics of the data may change over times, and retraining may be required at certain time intervals. For example, searches for patterns in business data may have changing requirements as time goes on so that new historical data needs to be used to retrain the network.

Another consideration is that the role and form of the explanation facility needs to be determined. If the neural network is combined with an expert system, the explanation facility of the expert system can deal with the higher level reasoning. However, at the neural network level, explanations based on the weight values are not always possible and additional software is needed to provide the logic to explain the result of the neural network.

9.3.1 Concept Phase

Taking a more detailed look at the concept phase of the development strategy, this activity needs to focus on understanding the problem to be addressed to see if a neural network solution is appropriate. Then, the development process is planned and certain decisions are made.

The neural network approach is particularly good for certain classes of applications. These include character recognition, signal processing, robotics and vision systems, and forecasting. Other types of applications should be avoided: traditional business applications such as accounts receivable that are done well with algorithmic techniques and applications that require deduction and sequential logic.

Current successful neural network applications have cited the following analysis of the situations faced by conventional techniques:

- Standard technology inadequate; not effective or too difficult to implement
- Qualitative or complex quantitative reasoning required
- Many highly-interdependent parameters are involved so that a clear understanding of the analysis and solution of a problem is difficult
- Data are intrinsically noisy or error-prone
- Project development time is short and the training time for the network is reasonable.

Another activity in the concept phase is the traditional step of gathering requirements about the problem to be solved and assessing resource needs and project risks. Constraints on the project include time and equipment resources; type, cost, accuracy, and timing of data; project, training, and operation times. A careful analysis of these aspects will ensure that, if proves to be feasible and desirable, the project will have a good chance of success.

Paradigm Selection

In this step, a network architecture (structure) and learning method are chosen. The characteristics of the application need to be analyzed and compared with the capabilities of different paradigms. Although ideally one would consider all possible alternatives, the availability of a particular development shell or the capabilities of development personnel may strongly influence the choice of the type of neural network to be used.

Areas of concern in paradigm selection include network size, nature of inputs and outputs, memory mechanism, type of training employed, and time constraints for routine operations of the working system. Examples of capabilities of neural network capabilities are show in the table:

Paradigm	Training Method	Training Time	Execution Time
Backpropagation	supervised	slow	fast
ART2	unsupervised	fast	medium
Kohonen	unsupervised	medium	fast
Hopfield	supervised	fast	medium
Boltzmann	supervised	slow	slow

An important consideration is the particular number of neurons and the number of layers to be used. The application problem needs to be analyzed carefully to see if the number of nodes required to represent the input and produce useful output is feasible for the available technologies. Within the group of current techniques, some are intended for smaller networks (a few hundred nodes) and others for on the order of 1,000 nodes. For example, the Boltzmann machine is a smaller network and backpropagation can be used for larger ones.

Neural network models vary in the type of output they can produce:

- Classification -- these applications generally require discrete outputs that specify to which of several categories a given input corresponds. Two or more output nodes may be required with only two active at a time, depending on the input to the network. For example, an application for forecasting product demand may have output nodes that correspond to high, medium, and low.

- Pattern recognition -- this widespread use of neural networks requires several nodes, some or all of which can be active. The particular output nodes that are active represent the pattern that has been identified.

- Real-number -- these outputs represent values such as machine settings, dollars, and degrees. Application areas include financial planning and robotics.
- Optimization -- applications such as those found in operations research use matrices of numbers to represent things such as resources to be allocated. Neural network implementations use the weights to calculate energy functions that are minimized to find the best solution to a problem. Examples are the traveling salesman problem and scheduling.

The memory mechanism influences how the network will recall things from partial information. For example, humans are excellent at reconstructing information so that, for example, another person can be identified from only a few clues about parts of the face.

One type of memory in neural networks is called associative and are represented by single-layer, highly connected neural networks. With associative memory, input patterns are correlated with corresponding outputs, and like biological vision systems, visual space can be mapped onto memory storage areas in organized patterns. Applications range from target classification to trend analysis of financial data.

Autoassociative memory networks take partial information and reconstruct patterns that were memorized previously. Each pattern is associated with itself; i.e., a clean version of itself is stored so that the exact pattern can be recalled even though future inputs may be incomplete or garbled. The Hopfield network, Brain-state-in-a-box, and adaptive resonance theory (ART) are examples of neural network models using autoassociative memory. Heteroassociative memory translates one set of patterns into a possibly different set, as in the function of look-up tables or translation systems (e.g., tables of ASCII code with the corresponding symbols). Examples of specific neural network models are BAM and Kohonen feature maps.

The training method depends on the type of data available and time constraints. The nature of the data may indicate <u>supervised learning</u>; i.e, the outputs associated with given inputs are well known. In that case, the application may need the neural network for the rapid production of desired outputs from future inputs. A set of inputs with corresponding correct outputs known from past experience are required for teaching the network how to respond.

In cases where classifications of inputs are uncertain, <u>unsupervised learning</u> may be needed in order to discover correlations in the data. With these techniques, the network creates its own classifications of input sets that are presented to the network. Certain parameters can be adjusted to make the network vary the number of categories it creates. The resulting outputs must then be inspected by humans to provide interpretations of their meanings.

Timing constraints are important for the successful use of a working neural network. The two factors are training time and operation time. The training time is the time for adjusting the weights by repeatedly presenting the training set until accurate results are produced. Different for the different types of neural network models and varies with the size of the network and the quality of the data. On the latter point, if the training data contains inconsistent information or does not cover all the possible cases evenly, the network can waste time trying to focus on the proper information to learn.

9.3.2 Application Design

During the concept phase, the application is analyzed for suitability for a neural network solution and, along with this feasibility study, an appropriate paradigm is selected. As more experience is gained with neural network development, more formal methods of risk analysis can be developed to assist project leaders in identifying and managing potential problems. If the prospects for the project are still favorable, we enter the design phase in order to plan specific steps and add detail to the development process.

Neural Network Design

The neural network is a system that needs to be designed at different levels: node, network, and training. We must determine the type of processing elements, the size and connectivity of the network layers, and the learning algorithm. The options are different among the various paradigms.

Neural networks are usually described in terms of a layer, which is a set nodes whose weights are modified as a unit. That is, during the training process, different groups of weights are affected at different times and each group would therefore be associated with the same layer. Although some people refer to the input and output nodes as a layer, we restrict are definition here to those nodes between them. Within a given layer, groups of nodes may have specialized purposes, e.g., representing a certain feature of the memory, and are in those cases referred to as slabs.

At the node level, the designer needs to address the types of inputs, the means of combining inputs, and the transfer function to be used. The means of combining is usually a simple summation over all the inputs of each weight times the corresponding activation level on that input line. Most neural networks in use today use a transfer function to determine if the node is to be activated and to determine the value of the output that goes on to another node.

The choice of transfer function, such as linear and sigmoid, is based on the types of inputs and outputs desires and on the learning algorithm to be used. With development shells, these choices would be determined already or presented to the developer for making a decision. For example, the backpropagation technique uses the sigmoidal transfer function because of the intrinsic nature of the training algorithm used.

At the network level, the designer must choose the number of layers, the size of each layer, the nature of the inputs and outputs, and the way each layer is to be connected. Network paradigms such as Hopfield, ART and Kohonen SOFM require by their nature one or two layers. Backpropagation, on the other hand, can have multiple ("hidden") layers, which perform the function of abstracting features form the inputs. Increasing the number of layers makes the network more powerful; however, the training time is greatly affected, both from the number of calculations the must be done and from the necessity of providing more examples in the training set. The hidden layers also make the network more difficult to understand in terms of explaining the reasons it produces its results. A good procedure is to start with one layer and experiment with the performance characteristics by adding additional layers. An important point to note is that if an operational network later is determined to require a different number of nodes, the modification will require essentially a complete redesign of the network and the input data.

In each layer, the number of nodes must be selected. The input nodes are chosen to represent the data sources adequately. This should proceed a thorough analysis of the data relevant to this application. Insufficient information in the data will lead to an inaccurate network and the presence of unnecessary data adds to training time and may also give inferior results.

In analyzing all the possible data sources relevant to the application, remove unnecessary and unreliable data and rule out data that are impractical or uneconomical to collect. Look for ways to preprocess data into efficient forms that will lead to desirable training times. For example, ratios of numbers are often more useful than raw data because information has been added that would otherwise have to be discovered by the network.

Next, the data must be put into the appropriate format for input to the type of neural network model being used. For example, data with continuous values are sampled at certain time intervals and input to the network. One node would be used for each source of continuous-valued data in the set.

If the type of network being used accepts only binary input, the data must be arranged as a set of numbers (a vector) with each element assigned to a particular input node. The order in which the elements are arranged in the vector can be important for the accuracy of and time required for training. For example, certain elements grouped together in the vector may represent a certain aspect, or feature, of the data and arranging them together will make it easier for the network to recognize that feature. Conversely, randomly arranging the elements will require more work and training time for the network to recognize features.

Some data are natural to represent as binary inputs. A series of attributes such as symptoms of a disease could use a 1/0 value to represent the presence or absence of that symptom. In other cases, where continuous values are involved, the values can be broken into ranges such as 0-3, 4-7, 8-11, etc. so that a node, or combination of nodes, with inputs 0/1 represent the continuous value. Patterns can be represented by a series of binary nodes. As shown in the figure above, characters can be represented as a vector whose elements represent the different pixels that make up the character.

As mentioned earlier, the number of nodes affects the training time and the performance of the network. On the latter point, too many nodes leads to a table-lookup effect in which the network rigidly maps the training set inputs to outputs. In this case, the network loses the ability of generalization that gives neural network technology one of its attractive advantages. With too many nodes, the network memorizes data rather than learning features of the data, gleaning basic principles for later use.

During the training process, statistics should be generated to gauge the progress of the training. At first, the percentage of correct responses to input data will be unacceptable, but improve as training goes on. In practice, the trained network may not be perfect, especially for large problems. In this case, a predetermined tolerance should have been set and used as a way to decide when to stop the training. During the design phase of a new system, the numbers of nodes and layers will usually be varied to come up with the best architecture before accepting the neural network for routine use in an application.

The system designer must determine how the results at the output modes will be used. Categorization applications will have a single node active to indicate to which of several categories an input corresponds. For pattern outputs, enough nodes are needed to represent the pattern and various nodes will be active for any given case. For example, the number nodes must be sufficient to represent each pixel and give the needed detail. In either case, post processing must be done to convey the neural network result to another system or to provide a user interface.

Various parameters are associated with each neural network paradigm, and they are opportunities for refining the system design. The number and type of parameters are peculiar to the paradigm, and a development shell makes it easy for the user to experiment with the settings and optimize the design.

Data Collection

The process starts with the collection of data to be used for training and testing. As described above, an important consideration is that the particular problem is amenable to neural network solution and that adequate data exist and can be obtained. Training data must be identified, and a plan must be made for testing the network performance.

The first two steps in the process involve collecting data and separating it into a training set and a test set. These tasks are based on a thorough analysis of the application so that the problem is well-bounded and the functionality of the system and the neural network's context are well-understood.

In conjunction with a domain expert, the developer must identify and clarify data relevant to the problem. This means formulating and conceptualizing the task in a data-oriented way that will be amenable to a neural network solution. Textual descriptions need to be reformulated to allow the inherent base of knowledge to be described numerically as vectors The developer needs to check for biases due to the particular way the data is represented and normalized. Another consideration is the stability of the input and the extent to which environmental conditions might require changes in the number of input nodes to the neural network or frequent changes in the training data. At this point, difficulty in expressing the data in the form needed for a neural network might lead to cancellation of the project.

In collecting and preparing the data, the anticipated structure of the neural network determines the data type, such as binary or continuous. High-quality data collection requires care to minimize ambiguity, errors, and randomness in the data. The data should be collected to cover the widest range of the problem domain and cover not only routine operations, but exceptions and conditions at the boundaries of the problem domain.

Another task is to confirm reliability by using multiple sources of data; however, ambiguities will have to be resolved. In general, the more data we use, the better as long as quality is not sacrificed. Larger data sets will increase the processing times during training, but better data will improve the accuracy of the training and could lead to faster convergence to a good set of weights.

Current neural network models have various parameter available for tuning the network to the desired performance level. Part of the process is the initialization of the network weights and parameters, followed by modification of the parameters as performance feedback is received. In many cases, the initial values are important for determining the efficiency and time length of the training.

Development Environment Selection

The last part of the design phase is matching the application description and design to an appropriate development environment. Ideally, all possible tools and techniques would be considered so that a thorough analysis of the application would indicate the best approach to use. In reality, various constraints will limit the possibilities, but to the extent possible this step should be done with objectivity to avoid using the wrong tool for the job.

As part of the design phase, the development environment selection should be a systematic analysis and give an accurate appraisal of the costs and time involved with various choices. Thus, at this stage, a poorly-chosen project could still be modified or canceled before the investment is made in a particular development tool.

One of the constraints in choosing the development environment is obviously the cost. The size of the project and the importance to the organization will be decision factors that could justify the costs. Another consideration is the familiarity of developers with tools used in previous projects. The cost in time and money of retraining personnel are important.

9.4 Analysis of Case Studies

The hybrid systems presented in Chapters 4-8 are examples of loosely and tightly coupled systems. They also represent a range of development tools and topics.

The LAM system for window glazing design is made up of independent modules that communicate via a common fact file that passes computational results and control parameters form one module to the next. The rule-based system can pass a structured query to the neural network and can use the results back form the neural network to make further inferences. The neural network takes the place of several rules by interpreting the data on solar and sound properties.

The systems for nuclear plant monitoring system and for multiple target recognition are tightly coupled systems linked via the HyperTalk commands. Their rules make use of neural network interpretations of detector data and sensor data. In the first case, the neural network determines plant operating states including normal and off-normal conditions. The expert system prediction knowledge to provide plant operators with intelligent recommendations. In multiple target recognition, the neural network is a classifier learning spatiotemporal attributes of target trajectories. The neural network offers robust pattern recognition performance, fault tolerance, and parallel processing capabilities. The expert systems provide symbolic processing and the overall decision making, management, and coordination functions.

The InFuse system is a tightly coupled system that uses neural networks to interpret textural descriptors obtained from imagery. The expert system verifies the neural network decisions using data from a spatial database and domain specific knowledge about sensor behavior.

The chemical tank control hybrid system demonstrates the use of the Conncert development concept that facilitates the embedding of neural networks in software systems. The network takes as input the history of PID controller input values and tank pH values and outputs expected pH values for later time steps. The expert system avoids over control by deciding when predicted outcomes are to be actually included in the controlling equation. The expert system uses a combination of temporal pattern matching and rules.

These cases provide detailed descriptions of working hybrid neural network and expert systems. The information about the applications and the data used in them allows a good understanding of the problems and opportunities involved in the development of hybrid systems.

9.5 Summary

This chapter has reviewed a number of ideas about the development of hybrid neural network and expert systems. Because hybrid systems is a new area of research and development, more elaborate and alternative methodologies and techniques are yet to be discovered.

The present development processes borrow heavily from conventional software development methodologies, and those principles should certainly be followed when the hybrid system is part of a larger software development project. However, as described above, certain additional considerations are imposed by the use of intelligent technologies.

As more experience is gained by hybrid system researchers and developers, new and better guidelines should emerge for the effective and efficient development of hybrid systems. Research efforts like those described in Chapter 2 will produce new techniques and tools, and the publication and analysis of further examples and case studies will lead to better insights into the ways to combine neural networks and expert systems for practical applications.

Chapter 10

Tools and Development Systems

The availability of neural network shells has allowed rapidly-increasing development of neural network applications, just as was the case historically for expert systems. Now that hybrid system development tools are starting to emerge, a similar surge in applications should occur.

10.1 Hybrid System Software

Most neural network development tools can support several network paradigms. In addition to the standard products, some special tools are spreadsheet-based, and other products are designed to work with expert systems as hybrid development products (see, for example, Table 10.1)

As with expert systems, the advantage of the shell is that extensive programming experience is not required to produce useful systems. Like any other development tool, the user of neural network tools is constrained by the configuration of the tool. Therefore, builders may prefer to use programming languages such as "C", or use spreadsheets to program the model and execute the calculations.

Development tools and environments for hybrid systems are just now starting to appear. However, most commonly individual expert system and neural network tools are linked in a loose-coupling mode (see Table 10.2). This allows a wide choice from the many products available for the two techniques individually. Alternatively, when programming expertise is available source code for expert systems and neural network modules can be developed or reused with necessary linking software developed as needed for the particular application. Currently common is development in C, and possibly in an object-oriented language, with a graphical user interface.

Product	Vendor
Brainmaker Professional	California Scientific Software
ExploreNet, NeuroSoft	Hecht-Nielsen Neurocomputer Corp.
MacBrain	Neurix
Nestor Development System	Nestor Corp.
NeuralWorks Professional	NeuralWare
NeuroShell, NeuroWindows	Ward Systems Group

Table 10.1 Examples of Neural Computing Tools

Some vendors are addressing the interface problem specifically to facilitate the linkage between expert systems and neural networks. NueX, by Charles River Analytics, Inc., is the first commercial system for developing hybrid neural network and expert systems. Gensym, Inc. has recently offered the NeurOn-Line tool for use in their G2 expert system development environment. Otherwise, various shells can be linked together, as for example the following:

Expert System Tools	Neural Network Tools
Exsys	NeuroShell
Level5	BrainMaker Professional
Nexpert Object	NeuralWorks Professional

Table 10.2 Examples of individual shells that can be linked (in any combination above) to create hybrid systems

Even when individual shells are being used for the neural network and expert system functions, some programming or database work needs to be done (see, for example, Labate and Medsker, 1993). Integration code may be needed to communicate intermediate results between the (possibly several) neural network and expert system components. Programming is also needed to prepare data for input to the neural networks and for possible training and retraining procedures.

File manipulation may be a major task and an obstacle to effective development of hybrid systems for some applications and shells.

Eventually, better development environments will be available for hybrid systems. As discussed in Chapter 2, research issues still need to be addressed to find the best representations and architectures for linking symbolic and subsymbolic systems. Techniques and tools need to be available to develop a variety of systems with less per application development. The Conncert tool, described in Chapter 2, is an example of a more fundamental and generic method (Wilson and Hendler, 1993).

10.2 Neural Network Hardware

Most of the current neural network implementations involve simulations on conventional sequential computers. However, advances in hardware technology will greatly enhance the performance of future neural network systems by exploiting the inherent advantage of massively parallel processing. Hardware improvements will meet higher requirements for memory and processing speed for shorter training times of larger networks.

Simulating a neural network means mathematically defining the nodes and weights assigned to it. So instead of using one CPU for each neuron, we use one CPU for all of the neurons. This simulation may take long processing time.

Each processing element computes node outputs from the weights and input signals from other processors. Together, the network of neurons can store information that may be recalled in order to interpret and classify future inputs to the network.

10.2.1 Traditional hardware solutions

In order to reduce the computational work of a neural network (it can be hundreds of thousands of manipulations) when the work is done on regular computers, one can use one of three approaches:

a. Faster machines. For example, a machine with Intel 80486 processor supplemented by a math coprocessor can expedite work, but not too much. (e.g. between 2 to 10 times faster).

b. Neural Chips. Most of today's chips can execute computations very fast but they do not train, so it is necessary to train off the chip. Until this problem is corrected, some developers are using acceleration boards. Chips can either be analog (e.g. Intel 80170 Electronically Trainable Neural network) or a digital chip or even an optical chip (see Caudill, 1991 for details). Most neural chips are still in the development stage.

c. Acceleration Boards are dedicated processors, similar to math coprocessors, that can be added to work very fast. Examples are Brainmaker Accelerator Board (from California Scientific Software), Balboa/860 boards (from HNC) and NeuroBoard (from Ward Systems Group).

10.2.2 Optical computing

Optical computers are being developed that use light beams instead of electrical currents (see Caudill and Butler, 1990 and Nelson and Illingsworth, 1991). For neural network applications, optical multiplication is performed using lenses and mirrors, and the optical multiplication of a matrix and a vector can be done extremely fast. The matrix represents the neural network weights and the vector values represent the strengths of the inputs.

With optical computers, the strength of an input signal is encoded as the intensity of a laser beam. neural networks require a laser for each input node of the and the intensities correspond to the strength of the input signal. The first lens spreads the beam so that it covers a rectangular area. This light then passes through a spectral light modulator (SLM), which selectively absorbs light in different parts of the rectangular area. This represents the different neural network weights. With the light from all the input lasers treated this way, the final intensity at the photoreceptors represents the matrix multiplication of the weights and the input vector. Fortunately, the SLM device can change its transparency pattern in response to an applied voltage. This corresponds to the weights being adjusted in response to the feedback required for the neural network to learn.

Optical computing is still a research area with some problems yet to be solved. For one, the physical size of the current devices with their lasers and bases, needs to be reduced. On-chip lasers may eventually solve this problem. Also, the systems are fragile and sensitive to alignment problems. Research is also needed to find better SLM's with the properties needed to represent more effectively the weight matrix. Good progress on optical neurocomputers is being made; however, commercial use is years away.

10.3 The NueX Development Environment

NueX is a commercially-available environment for developing hybrid neural network and expert systems. NueX was used in the cases presented in Chapters 5 and 8 and is described here in more detail.

The architecture of NueX, shown in Figures 10.1 and 10.2, is an integrated approach to structuring, representing, and processing domain knowledge. The Rule Base provides the prescriptions for using facts and assertions, which are stored in the Data Base. The Inference Engine uses forward reasoning and heuristic search strategies to reach conclusions pertaining to the problem at hand.

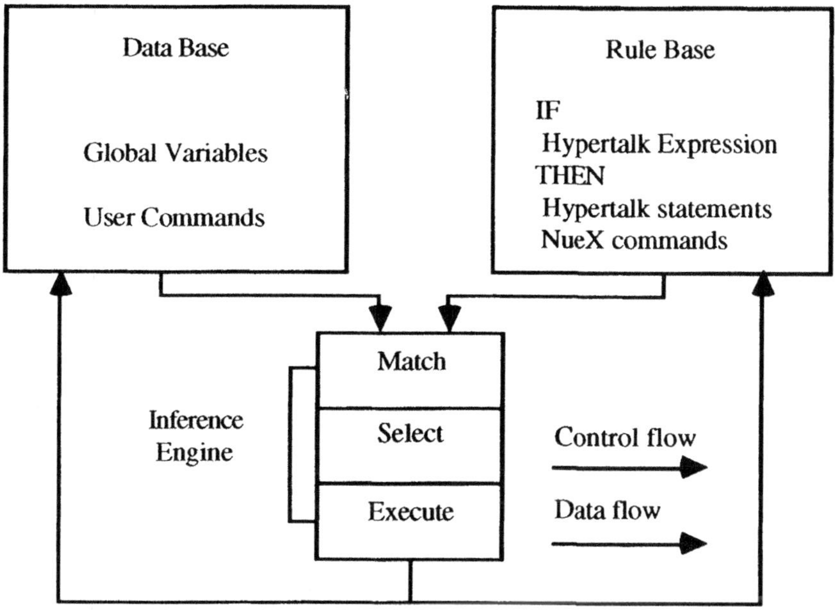

Figure 10.1 Architecture of the NueX hybrid system development environment

The Figures 10.3-10.6 show some of the features of NueX that are useful in developing hybrid systems applications. First, because NueX uses hypercard, the output interface can be more elaborate than simply printing results. In this case, an application scenario can be portrayed and changed by clicking on items in a picture.

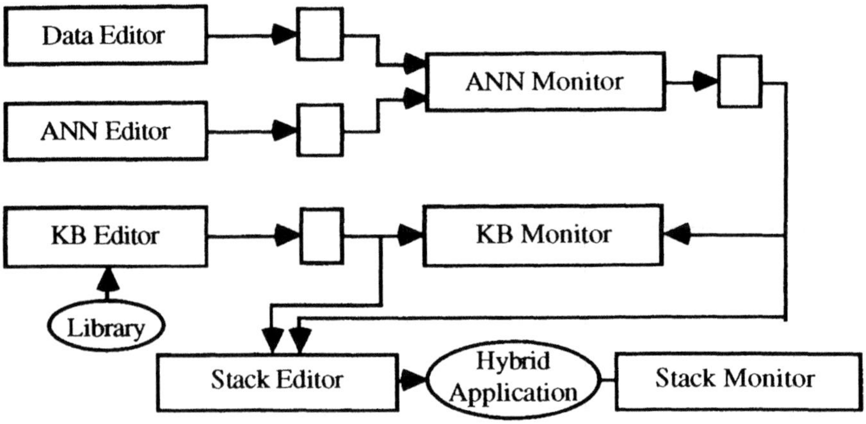

Figure 10.2 Components of the NueX development environment

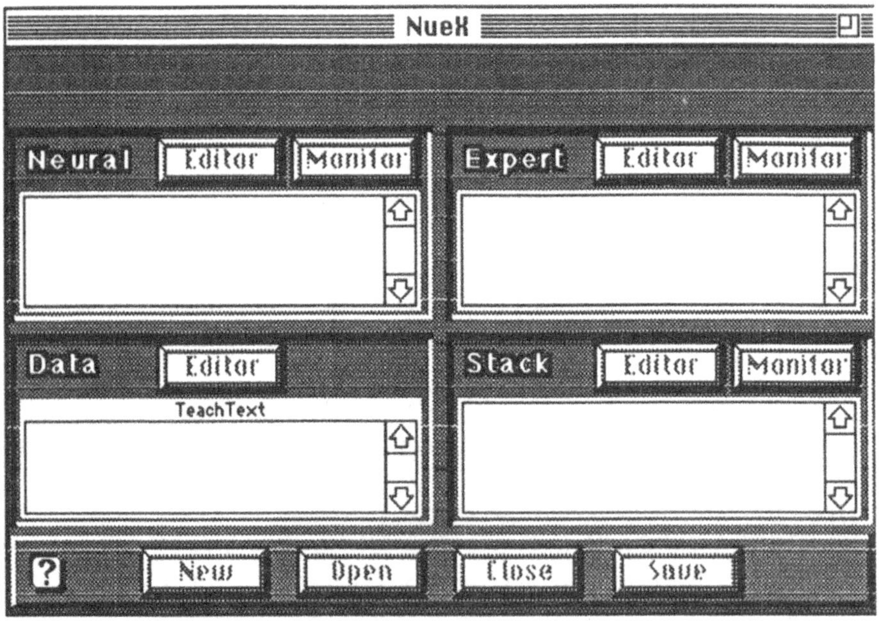

Figure 10.3 Initial screen in the NueX development environment

In NueX, the Data Base is the list of HyperTalk global variables defined by the developer using the KB Editor (see Figure 10.2). The operation of the inference engine can be observed using the KB Monitor. The ANN Editor provides a graphical means of specifying the design of and the parameters for the neural networks.

Figure 10.3 shows the initial screen presented to the developer. The application stack and its associated knowledge base and neural network can be selected, if already created, or a new application can be started. Figures 10.4 is an examples of a rule, with distinct sections for the IF and THEN parts. Library commands can be invoked to produce links to the neural network and to specify HyperText interface instructions.

In the neural network side of NueX, the architecture is specified graphically as shown in Figure 10.5. Via pull-down menus and sidebars, nodes and layers can be designed in the configuration desired. The training of the neural network can be monitored, as shown in Figure 10.6, to check the progress and look for changes to make in the parameters in subsequent retraining attempts.

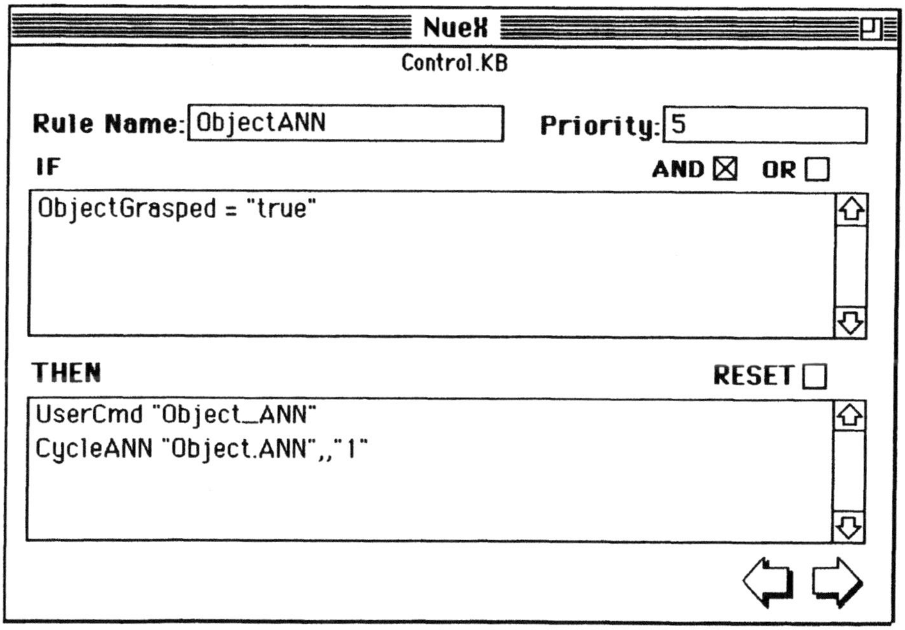

Figure 10.4 An example of rule entry in the NueX environment

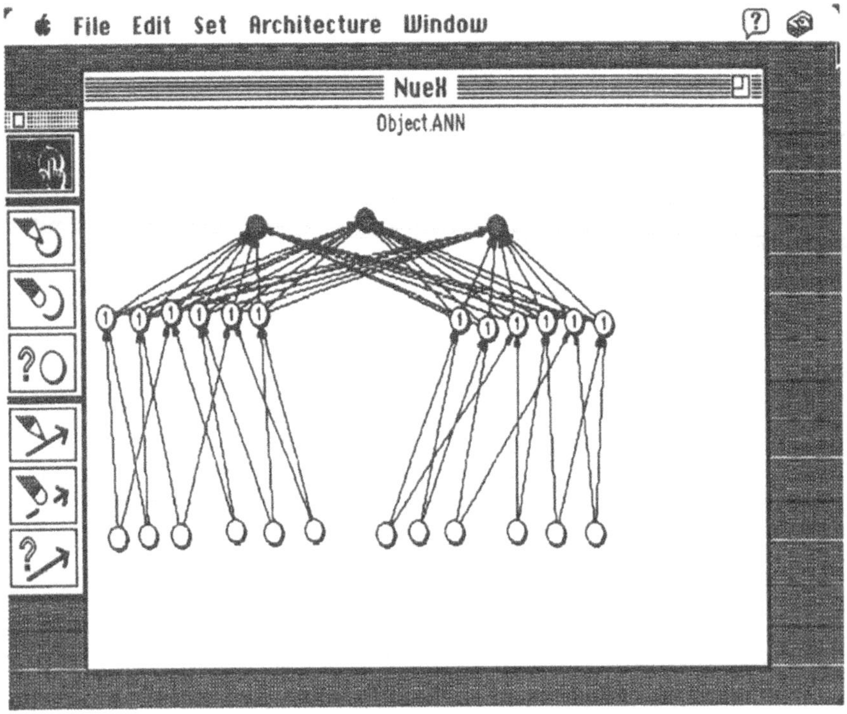

Figure 10.5 An example of network design entry with NueX.

The NueX environment is one of the first commercially available environments for developing hybrid neural network and expert systems. The discussion in this section illustrates the nature of NueX and shows the overall strategy for such environments. Other products will appear as interest in hybrid systems continues to grow.

10.4 Other Development Environments

10.4.1 Carnegie Group, Inc.

The Carnegie Group has recently announced an expert system and neural network hybrid technology designed for diagnostic and repair problems that involve signal data as well as domain knowledge. An initial use of their system in manufacturing applications involves diagnosis of chatter in cold-rolling mills.

They use NeuralWorks Professional from NeuralWare, Inc. for the neural network modules and their own TestBench expert system tool.

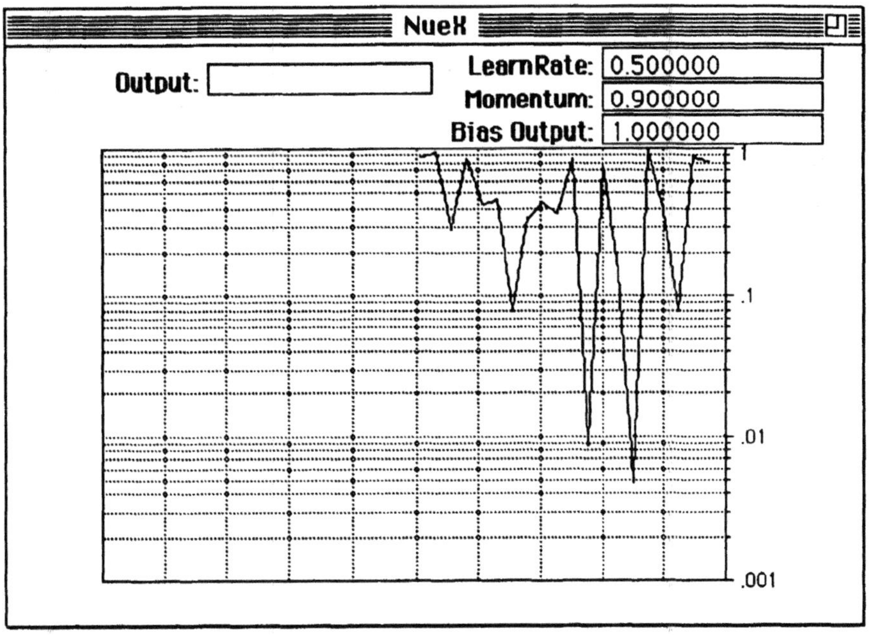

Figure 10.6 An example of monitoring training in the NueX environment

10.4.2 Gensym Corporation

Gensym recently offered a NeurOn-Line to be used on their G2 real-time expert system development environment, which is intended especially for applications such as those in Computer Integrated Manufacturing (CIM).

G2 is a development and delivery system that allows real-time execution, heuristic and procedural reasoning, dynamic simulation, graphical interface, and database capabilities. G2 provides a structured natural language editor and an object-oriented graphical interface, linkage to FORTRAN and C modules, explanation system, and fuzzy logic handling.

As shown in Figure 10.7, G2 includes several products useful for the development of integrated systems. The G2 Diagnostic Assistant (GDA) is a graphical, object-oriented programming system for diagnostic modules. The

Dynamic Scheduling Package (DSP) allows constraint-based scheduling system development. Telewindows allows many on-line users to interact with G2 applications via the G2 Network. The latter can also connect to other systems such as sensors, databases, and controllers.

NeurOn-Line has been used effectively for complex, low-level pattern recognition in applications involving acoustic or vision systems for monitoring and inspecting. The neural network output is processed by G2 for further conclusions and decision making.

Examples of hybrid systems developed with G2 are SkilTracer, which diagnoses problems in CIM. The neural network learns patterns for robot arm motion and G2 monitors operations and supervised retraining when needed. Another application involving NeurOn-Line oversees the construction and training of neural networks.

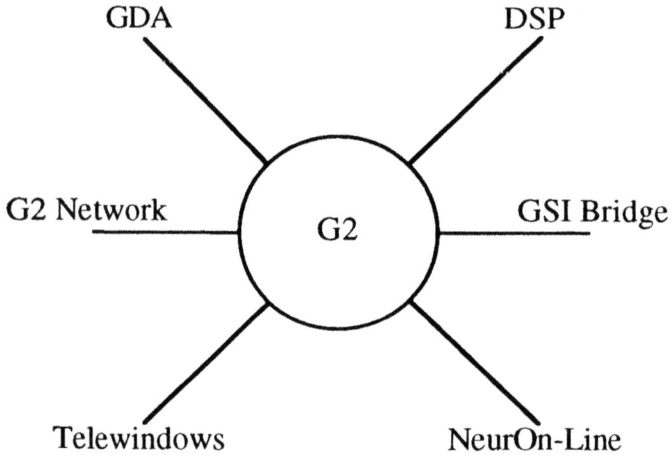

Figure 10.7 The G2 environment and family of products

10.5 Summary

The proliferation of expert system shells and the recent availability of neural network shells and tools have enabled the rapid growth in the number of hybrid systems. These systems tend to be loosely coupled, using file transfer to communicate between the shell-based systems.

With the prospects for useful hybrid systems have come commercial products to facilitate the linkage between expert system and neural network subsystems. However, the success of hybrid systems is leading to new development environments that focus upon and facilitate the production of hybrid systems as the primary objectives.

Along with the progress on software tools, the economical availability of hardware for parallel processing will increase the attractiveness of neural network solutions and therefore increase the advantages of hybrid systems. Both the hardware and software advances will take advantage of the improved approaches and new techniques that will emerge form the research and development efforts described in Chapter 2.

Chapter 11

Summary and the Future of Hybrid Neural Network and Expert Systems

11.1 Introduction

Expert systems perform reasoning using pre-established rules for a well-defined and narrow domain. They combine knowledge bases of rules and domain-specific facts with information from clients or users about specific instances of problems in the knowledge domains of the expert systems. Ideally, reasoning can be explained and the knowledge bases easily modified, independent of the inference engine, as new rules become know.

Expert systems are especially good for closed-system applications for which inputs are literal and precise, leading to logical outputs. They are especially useful for interacting with the client/user to define a specific problem and bring in facts peculiar to the problem being solved. ES reasons using established facts and pre-established rules.

A major limitation of the expert system approach arises from the fact that experts do not always think in terms of rules. Thus, an expert system does not in such cases mimic the actual reasoning process of human experts. For example, credit application evaluation involves aspects such as human factors and organizational issues; theses are difficult to be captured in rules, but appear as text in the application processing approval.

Neural networks rely on training data to "program" the systems. Particular applications are developed by establishing an appropriate training set that allows the system to learn and generalize for operation on future input data. Inputs that exactly match the training data are recognized and identified, while new data (or incomplete and noisy versions of the training data) can be put into the closest matches to patterns recognized by the system.

For stable applications with well-defined rules, expert systems can be easily developed to provide good performance. Furthermore, most development software packages allow the creation of explanations to help the user understand questions being asked or conclusions and reasoning processes. In contrast, in ANN, knowledge is represented as numeric weights, and therefore, the rules and the reasoning process are not readily explainable.

Neural networks can be preferable to expert systems when rules are not known, either because the topic is too complex or no human expert is available. If training data can be generated, the system may be able to learn enough information to function as well as, or better than, an expert system. This approach also has the benefit of easy maintenance. That is, modifications are exercised by retraining with an updated data set, thus eliminating programming changes and rule reconstruction. The data-driven property of neural networks allows adjustment of changing environments and events. Expert systems are knowledge driven so changes can be made only if the knowledge is being changed. Another advantage of the neural network implementation is the speed of operation after the network is trained; the natural use of parallel systems and neural chips enhances this aspect dramatically.

The differences between expert systems and neural networks require care in choosing the appropriate technology for a particular application. The two technologies may represent complementary or competing approaches. ANN can be the best solution for some of the problems that have proven difficult for expert system developers. The special characteristics of the technologies create opportunities for integrated systems that allow system developers to address problems not amendable to either approach alone.

Guidelines for choosing appropriate applications for hybrid ES and ANN are starting to emerge. Some important conditions in which neural networks may be useful are as follows:

Inadequate knowledge bases. An expert may not be available or be affordable; rules be very difficult to formulate; the rule-based approach may not be an appropriate representation for the domain. Adding an ANN as a front-end may help.

Volatile knowledge bases. Rules and facts may be frequently modified; knowledge in rules may be evolutionary, dependent on human experience in a new domain. ANN, if feasible, handles such changes better.

Data-intensive systems. Applications involving high rates of data input, requiring rapid processing; feedback data for control systems; ambiguous, noisy, or error-prone inputs requiring interpolation; interactions involving vision and speech subsystems, are all candidate for ANN support.

Parallel hardware. Availability of parallel implementations adds additional advantages to the neural computing approach.

11.2 Other Intelligent Technologies

Expert systems and neural networks are particularly good technologies to integrate because of their complementary natures. However, either or both can be combined with other intelligent subsystems to address deficiencies in current solutions to real-world problems.

Fuzzy Systems are based on fuzzy set theory and associated techniques pioneered by Lotfi Zadeh (Zadeh, 1965). A goal of this approach is to mimic an aspect of human reasoning by doing approximate reasoning. Thus, such systems are excellent components of complex hybrid systems. Fuzzy systems can provide an interface that seems more like the interactions people experience with each other.

Applications of fuzzy logic have recently appeared in consumer products, including autofocusing cameras, washing machines, and microwaves. Also, special software packages for developing fuzzy system are becoming available and should markedly increase the number of such system in use.

Fuzzy systems are less precise than conventional systems but are more like our everyday experience as human decision makers. We tend to talk in fuzzy terms such as "tall," "large," and "rarely." Fuzzy systems allow users to give input in these imprecise terms and use them to give fuzzy or precise advice. The internal logic system is designed to deal with fuzzy terms and give useful conclusions. This technique can broaden the usefulness of expert systems, allowing operation in gray areas where precise values may not be know or may not be necessary for drawing conclusions.

Object-oriented Techniques have recently become very popular as a general approach to automated problem-solving, including the development of expert systems. The relationship with expert systems is very strong because of the connection with the knowledge representation techniques using frames and semantic networks.

The object-oriented approach to programming, and the design of systems in general, is a nonconventional way of modeling situations. The approach is gaining much attention as a useful way to describe data, relationships, and processes that are associated with reasoning.

In object-oriented techniques, the application area is viewed as a set of objects with associated attributes, including procedures and processes. Interaction between objects is accomplished by sending and receiving messages. The abilities to define objects and pass messages are capabilities that can be added to a programming language or a design tool.

The mechanism called inheritance is the capability of the design or development tool to store hierarchical information that can be accessed efficiently when needed. Object-oriented techniques thus foster modular systems that are easier to design and maintain. Expert system shells are starting to include object-oriented features, allowing easier development of hybrid systems.

Case-based Reasoning (CBR) is a relatively new field based on the idea of making use of solutions to previous problems for solving new ones. Development systems are now available to help analyze historical information and put it into a form that is useful for subsequent problem solving.

CBR was conceived as a technique that might be similar to an aspect of human reasoning; i.e., we tend to refer to past experiences for guidance in solving current problems. A CBR tool assists the developer in storing solutions to previous problems and determine differences from new problems. This involves finding many relevant cases, finding the best matches to the current problem, and analyzing the input characteristics. The differences compared to the problem at hand are then used to modify the old solutions.

CBR can be used by itself or as part of a hybrid intelligent or conventional system. CBR is especially indicated when rules for the knowledge domain are difficult to discern and/or the number and complexity of rules is too large for the normal knowledge acquisition process . Some domains have particularly rich histories of precedents. Examples are legal applications and medical and other diagnostic systems.

The **Genetic Algorithm** (GA) technique represents intelligent systems by mimicking biological systems that self-organize and adapt to their environments. Modeling after concepts of biological evolution, GA systems use feedback from interaction with the environment. Weak or negative feedback causes certain possible solutions to a problem to drop out of consideration, while other feedback causes better candidate solutions to survive.

GA can also be considered as a search technique that is an alternative to those in traditional AI. GA can find a local optimum solution that is adequate, and thereby not having to do an exhaustive search. This is good for complex situations in which the search space would be prohibitively large for AI techniques.

GA is an interesting technique that is the receiving considerable attention in research and development labs. An advantage over expert systems is the aspect of learning from experience and the use of information to identify inferior and preferred solutions. Thus, GA components can enrich a hybrid system and raise the level of intelligence.

A theme that can be found in these alternatives is the attempt to make up for deficiencies in the expert system and other AI approaches. Another aspect is the need to add additional features associated with human intelligence such as learning and the ability to generalize from current knowledge.

11.3 Neurocontrol

Neurocontrol is a new field that could lead to the understanding of the human brain and mind (Werbos, 1993) while leading to important new applications that have not been possible with AI and neural network techniques. Current neural networks are simple functions that could be combined for more complex brain-like systems.

In this approach, the brain is considered, as a whole system, to be a neurocontroller subject to the mathematics of neurocontrol. Werbos (1993) explains, though, that this goes beyond applied conventional control theory and uses a broader definition to model brain-style intelligence.

Werbos has identified one class of design, optimizing design, that is plausible for modeling brain-style behavior. Optimizing design attempts to maximize a measure of success or performance (utility) over time. Optimization over time is a generic task that goes beyond, and can make use of, artificial neural network subsystems that learn to approximate specific limited functions. We need to learn how to combine these subsystems into complex systems that learn for to optimize future results by making initial decisions.

For true dynamic optimization, only two useful families of design are available: the backpropagation of utility and the adaptive critic design, in which an artificial neural network component serves as a "critic" or evaluation function (Werbos, 1993). Adaptive critic designs are preferred because they are extremely diverse, handling complex as well as simple situations.

Optimal neurocontrol has been applied successfully in a number of areas n which current actions must be chosen to optimize later results. Examples are the use of neurocontrol in the National Aerospace Plane (NASP), controllers to reduce waste in industrial processes and automobiles, and robotic and manufacturing systems.

"Dynamic optimization designs have experienced tremendous practical success and theoretical progress in the past few years ... the links to the brain and the mind are surprisingly rich and detailed." (Werbos, 1993) Future hybrid systems should make use of the likely advances in neurocontrol to produce useful systems that provide higher levels of intelligence.

11.4 Summary

Whether through brain-like systems or simpler combinations of intelligent technologies, more complex systems integrating intelligent technologies are increasingly feasible and desirable for solving practical problems. Both from the AI and neural network communities, the more difficult problems are being addresses with combinations of technologies.

The interest in hybrid systems is apparent in the number of tutorials and workshops sponsored by the International Neural Network Society (INNS), the American Association for Artificial Intelligence (AAAI), the International Association of Knowledge Engineers (IAKE), and other organizations. Those groups, as well ones representing specific engineering and application areas, are forming special interest groups to share information and promote awareness of advances and opportunities in the area of hybrid systems.

In the next few years, we should see exciting advances in the theoretical understanding and the practical use of intelligent technologies for hybrid systems. The availability of more useful development environments will make the technologies and techniques more accessible to larger numbers of people. The exploration of broader areas of application will uncover new research problems and provide insight and information for the construction of models and guidelines for the development of future systems. In this regard, progress in hybrid systems will be enhanced through the sharing of information on successful integration projects and through the analysis of how and why combinations of intelligent technologies can produce more powerful and effective systems than can be achieved through the use of any of the technologies alone. And we can continue to look inward at the brain, which provides the best model yet of integrated specialized subsystems optimized for intelligent behavior.

References

AAAI (1990), "Integrating Symbolic and Neural Processes," Workshop Notes from the AAAI Eighth National Conference on Artificial Intelligence, Boston, 1990.

AAMA (1984), AAMA Bulletin CW-12, "Structural Properties of Glass," American Architectural Manufacturers Association, Des Plaines, IL

AAMA (1987), AAMA Bulletin GDSG-1, "Glass Design for Sloped Glazing," American Architectural Manufacturers Association, Des Plaines, IL.

AAMA (1988), ANSI/AAMA 101-88, "Voluntary Specifications for Aluminum Prime Windows and Sliding Glass Doors," American Architectural Manufacturers Association, Des Plaines, IL.

Acar, L. and Ozguner, U. (1990), "Design of knowledge-rich hierarchical controllers for large functional systems." *IEEE Trans. on Systems, Man, and Cybernetics*, Vol. 20, No. 4.

Allen, J. F., and Miller, B. W. (1988), The rhetorical knowledge representation system: a user's manual. Technical Report TR 238, University of Rochester.

Allman, W. (1989), *Apprentices of Wonder*, Bantam Books, New York.

Antsaklis, P. J., Passino, K. M., and Wang, S. J. (1991), "An introduction to autonomous control system." *IEEE Control Systems*, pp .

AS (1989), Australian Standard AS 1288, "Glass in Buildings - Selection and Installation," Standards Association of Australia, North Sidney.

ASTM (1985a), ASTM Standard C 1036-85, "Specification for Flat Glass," Philadelphia: American Society of Testing Materials.

ASTM (1985b), ASTM Standard C 1048-85, "Specification for heat-treated flat glass - Kind HS, Kind FT, Coated and Uncoated Glass," American Society of Testing Materials, Philadelphia.

ASTM (1989), ASTM Standard E 1300-89, "Practice to Determine the Minimum Thickness of Annealed Glass Required to Resist a Specified Load," American Society of Testing Materials, Philadelphia.

Baffes, P. T. (1989), "NETS User's Guide," Software Technology Branch, Lyndon B. Johnson Space Center.

Bailey, D. L., Thompson, D. M, and Feinstein, J. L., (1988), "The practical side of neural networks." *PC AI*, vol Nov./Dec., pp. 33-36.

Bailey, D. L. and Thompson, D. M. (1990), "How to develop neural-network applications." *AI Expert*, vol 5, No. 9, pp. 34-41.

Barker, D. (1990), "New partners in the AI dance: neural networks and expert systems," *AI Week*, vol. 7, no. 9, pp. 1-6.

Barnden, J. (ed) (1991), *Advances in Connectionist and Neural Computation Theory*, Ablex.

Barnden, J. (1992), "Belief representation and connectionism," in [Sun, Bookman, and Shekhar, 1992] at pp. 106-114.

Bar-Shalom, Y. (1978), "Tracking methods in a multitarget environment," *IEEE Trans. on Auto.matic Control*, Vol. AC-23, No. 4, pp.

Barrilleaux, J. M. (1990), "A biologically motivated algorithm for image interpretation based on multi-pass multi-resolution techniques," *Proc. of the IJCNN*, San Diego, (June), pp. 813-818.

Beale, R., and Jackson, T. (1990), *Neural Computing*, Adam Hilger, Bristol, England.

Beason, W. L., and Barry, C. J. (1992), :Development of a new glass thickness selection procedure," *ASTM Standardization News*, November, pp. 62-65.

Beers, P. (1993), "The lasting effects of hurricane Andrew," *Glass Magazine*, vol. 43, no. 4, pp. 24-31.

Benachenhou, D., Cader, M., Szu, H., Medsker, L., Wittwer, C., and Garling, D. (1990), "Neural networks for computing invariant clustering of a large open set of DNA-PCR primers generated by a feature-knowledge based system." *Proc. IJCNN-90*, San Diego, CA, vol ii, pp. 83-89.

Bhatnagar, R., and Kanal, L. (1992), "Reasoning in uncertain domains: a survey and commentary," in *Encyclopedia of Computer Science and Technology*, Marcel Dekker Inc., New York.

Blake, A. and Zisserman, A. (1987), *Visual Reconstruction*, MIT Press, Cambridge, MA.

Bookman, L. A. (1992), "A two-tier framework for comprehension," in [Sun, Bookman, and Shekhar, 1992] at pp. 66-74.

Brachman, R. J., Gilbert, V. P., Levesque, H. J. (1985), "An essential hybrid reasoning system: knowledge and symbol level accounts of KRYPTON," *Prodeedings of the Ninth International Joint Conference on Artificial Intelligence*, Los Angeles, pp. 532-539.

Caglayan, A. K. and Allen, S. M. (1990), "A Neural Net Approach to Space Vehicle Guidance," *Proc. American Control Conference*, San Diego, CA.

Caglayan, A. K., Allen, S. M., and Edwards, S. J. (1989), "Hierarchical Damage Tolerant Controllers for Smart Structures," AFWAL-TR-89-3009 (March).

Caglayan, A. K., Allen, S. M. and Rahnamai, K. J. (1989), "Failure Detection, Isolation and Estimation for Reconfiguration of Aircraft Flight Control Systems Subjected to Actuator Failure and Surface Damage," Charles River Analytics Inc. WRDC-TR-89-3058 (June).

Calderone, I., and Melbourne, W. H. (1993), "Taking a closer look at glass strength," Society of Engineers, Australia, Department of Mechanical Engineering, Monash University, September.

Callan, J. P. (1990), "Automatic feature construction from domain information," in [AAAI, 1990] at p. 12.

Carpenter, G. and Grossberg, S. (1987), "A massively parallel architecture for a self-organizing neural pattern recognition machine," *Computer Vision, Graphics and Image Processing*, 37, 54-115.

CASYS, (1990), *Charles River Analytics Inc. CASYS User's Manual Version 2.083.*

Caudill, M. (1990), "Using neural nets: hybrid expert networks," *AI Expert*, vol 5, no. 11, pp. 49-54.

Caudill, M. (1990), Using neural nets. (Six Parts) *AI Expert*, December, April, June, July, September, December.

Caudill, M. (1990), Using neural nets: hybrid expert networks, *AI Expert*, (November), pp. 49-54.

Caudill, M, and Butler, C. (1990), *Naturally Intelligent Systems*, MIT Press, Cambridge, MA.

Caudill, M. (1990), "Using neural nets: hybrid expert networks," *AI Expert*, vol 5, no.11, pp. 49-54.

Caudill, M. (1991a), "Neural network training tips and techniques," AI Expert, vol 6, (January), pp. 56-61.

Caudill, M. (1991b), "Embedded neural networks," AI Expert, vol 6, (April), pp. 40-45.

Chang, C. B. and Tabaczynski, J. A. (1984), "Applications of state estimation to target tracking." *IEEE Trans. on Auto.matic Control*, Vol. AC-29, No. 2, pp.

Charniak, E. (1991), "Bayesian networks without tears," *AI Magazine*, Winter, pp. 50-63.

Chettri, S.R., Cromp, R.F., and Birmingham, M. (1992), "Design of neural networks for classification of remotely sensed imagery", in *Proc. Goddard Conference on Space Applications of Artificial Intelligence*, pp. 137-149.

CGSB (1989), CAN/CGSB-12.20-M National Standard of Canada "Structural Design of Glass for Buildings." Ottawa: Canadian General Standards Board, publication by Canadian Government Publishing Centre.

Cohen, M. E., and Hudson, D. L. (1992), "Integration of Neural Network

Techniques with approximate reasoning in knowledge-based systems," in [Kandel and Langholz, 1992] at pp. 72-87.

Cohn, A. G. (1989), "Taxonomic reasoning with many-sorted logics," *Artificial Intelligence Review*, vol 3, pp. 89-128.

Conlisk, P. (1992), "MSC/NASTRAN analysis of shear transfer in laminated architectural glass," *Proc. MSC 1992 World Users Conference*, May, session 1A, paper 2.

Cui, Y., Hendler, J., Su, H., and McAvoy, T. (1993), "Improving pid controllers with a neural network," Technical Report, Institute of Systems Research, University of Maryland, August.

Dalgliesh, A. (1989), "Canada Sets a New Standard." Glass Canada, June, pp. 22-23.

Dattatreya, G., and Kanal, L. (1985), "Decision trees in pattern recognition", in *Progress in Pattern Recognition 2*, eds: Kanal, L. and Rosenfeld, A., North-Holland Publishing Co., Amsterdam.

Daugman, J. G. (1988), "Complete discrete 2-d gabor transforms by neural networks for image analysis and compression," *IEEE Trans. Acoustics, Speech, and Signal Processing*, vol. 36, pp. 1169-1179.

Davis, K. (1988), "Diagnostic Expert System for the B1B," *IEEE AES Magazine*, (April).

Dolenga, J. (1993), "Hurricane-proofing South Florida," *Glass Magazine*, May, pp. 38-39.

ESDU (1971), ESDU Item Number 71013, "Elastic Direct Stresses and Deflections for Flat Rectangular Plates Under Uniformly Distributed Normal Pressure." (With 1987 amendment.) London, England: Engineering Sciences Data Unit, Ltd.

Fischler, M. A. *et al.* (1978), "The SRI road expert: an overview," *Proc. of Image Understanding Workshop*, (November), pp. 13-19.

Foss, R. V., and Droste, D. H. (1990), "An expert system for window glazing design," *ASHRAE Transactions*, vol 96, part 2, Atlanta, pp. 563-569.

Frisch, A. M. and Cohn, A. (1991), "Thoughts and afterthoughts on the 1988 workshop on principles of hybrid reasoning," *AI Magazine*, vol 11(5), pp. 77-83.

Frisch, A. M. and Scherl, R. B. (1991), "A bibliography on hybrid reasoning," *AI Magazine*, vol 11(5), pp. 84-87.

Fu, L. (1992), "A parallel approach for knowledge-based inference and learning," in [Kandel and Langholz, 1992] at pp. 88-103.

Gallant, S. I. (1988), "Connectionist expert systems," *Communications of the ACM*, vol 31, pp. 152-169.

Gallant, S. I. (1993), *Neural Network Learning and Expert Systems*, MIT Press, Cambridge, MA.

Garson, G. D. (1991), "Interpreting Neural-Network Connection Weights," AI Expert, vol 6, (April), pp. 47-51.

Glover, D.E. (1988), "An optical fourier/electronic neurocomputer automated inspection system." *Proc. of the IEEE Int. Conf. on Neural Networks*, pp. 569-576.

Glover, C. W. and Rao, N. (1990), "Hybrid Neural Network and Rule–Based Pattern Recognition System Capable of Self–Modification." *Applications of Artificial Intelligence VII*, Vol. 1293.

Gomi, T., and Nakamura, N. (1990), "The integration of symbolic and subsymbolic processes in the design of an autonomous vehicle," in [AAAI, 1990] at p. 18.

Gonsalves, R. G., and Caglayan, A. K. (1992), "A hybrid neural network/expert system approach to multiple target recognition," Charles River Analytics, Inc. Report No. R91121, Cambridge, MA, January.

Gutknecht, M., Pfeifer, R., and Stolze, M. (1991), "Cooperative hybrid systems." *Institut fur Informatik Technical Report,* Universitat Zurich.

Hall, L. O., and Romaniuk, S. G. (1992), "Performance issues of a hybrid symbolic connectionist learning algorithm," in [Kandel and Langholz, 1992] at pp. 106-133.

Hanson, M. A. and Brekke, R. L. (1988), "Workload management expert system - combining neural networks and rule-based programming in an operational application," *Proc. Instrument Society of America*, vol 24, pp. 1721-26.

Harston, C.T. (1990), "Applications of neural networks to robotics," in [Maren, Harston, and Pap, 1990] at pp. 381-385.

Hebb, D. O. (1949), *The Organization of Behavior*, Wiley, New York.

Hecht-Nielsen, R. (1990a), *NeuroComputing*, Addison-Wesley, Reading, MA.

Hecht-Nielsen, R. (1990b), "A theory of back propagation networks," *Proc. of the IJCNN*, Washington, DC, (January),1990, pp. XX-YY.

Hendler, J. (1989a), "On the need for hybrid systems," *Connection Science*, vol 1, no. 3, pp. 227-229.

Hendler, J. (1989b), "Marker-passing over microfeatures: towards a hybrid symbolic/connectionist model," *Cognitive Science*, vol 13, no. 1, pp. 79-106.

Hendler, J. (1991), "Developing hybrid symbolic/connectionist models," in [Barnden, 1991] at pp. 165-179.

Hendler, J., and Dickens, L. (1991), "Integrating neural network and expert reasoning: an example," *Proc. AISB Conf. on Developments of Biological Standardization*, Leeds, United Kingdom, 1991.

Honovar, V. (1990), "Toward generalized connectionist networks: an integration of symbolic and subsymbolic approaches to the design of intelligent systems," in [AAAI, 1990] at pp. 22-23.

Hopfield, J. (1982), "Neural Networks and Physical Systems with Emergent Collective Computational Abilities. *Proc. Natl. Acad. Sci. USA* vol 79, pp. 141-152.

Jones, D. and Franklin, S. P. (1990), "Choosing a Network: Matching the Architecture to the Application," in [Maren, Harston, and Pap, 1990] at pp. 219-231.

Kanal, L. (1972), "Interactive Pattern Analysis and Classification Systems - A Survey and Commentary", *Proc. of IEEE*, vol 60, no. 10, pp. 1200-1215.

Kanal, L., and Chandrasekaran, B. (1972), "Linguistic, Statistical and Mixed Models for Pattern Recognition", in *Frontiers of Pattern Recognition*, edr: M.S. Watanabe, Academic Press.

Kanal L. (1979), "Problem-Solving Models and Search Strategies for Pattern Recognition", *IEEE Trans. on Patt. Anal. and Mach. Intell.*, vol 1, no. 2, pp. 193-201.

Kanal, L. (1990), *Neural Networks and Statistical Pattern Recognition*, Sethi and A. Jain (eds.), Preface.

Kanal, L. and Raghavan, S. (1992), "Hybrid Systems - A key to intelligent pattern recognition," *Proc. of the IJCNN*, Baltimore, Part IV, pp. 177-183.

Kandel, A., and Langholz, G. (eds) (1992), *Hybrid Architectures for Intelligent Systems*, CRC Press, Boca Raton.

Keyes, J. (1991), "AI on a Chip," AI Expert, vol 6, (April), pp. 33-38.

Knaus, R. (1992), "Representing expert knowledge in neural nets," in [Kandel and Langholz, 1992] at pp 346-355.

Kohonen, T. (1984), *Self Organization and Associative Memory*, Springer-Verlag, Berlin.

Kohonen, T. (1988), "An Introduction to Neural Computing," *Neural Networks*, 1 (no. 1, 1988): 3-16.

Kosko, J (1989), "New Glass Specification Charts Developed," *Glass Magazine*, March, pp. 44-45.

Kuczewski, R. (1989), "TRW investigation of scan-scan correlation and multi-target tracking," *DARPA Neural Network Study*, Lincoln Lab Report No. 840 (March).

Kuncicky, D. C., Hruska, S. I., and Lacher, R. C. (1991), "The equivlaence of rule-based expert system and artificial neural netwrok inference," *International Journal of Expert Systems*, vol 4, no. 1, pp 281-297.

Kwasny, S. C., and Faisal, K. A. (1991), "Rule-based training of neural networks," *Expert Systems with Applicatons*, vol 2, no. 1, pp. 47-58.

Labate, F. and Medsker, L. (1993), "Employee skills analysis using a hybrid neural network and expert system," *Proceedings of the IEEE International Conference on Developing and Managing Intelligent System Projects*, Washington, DC, 1993, pp. 205-211.

Lacher, R. C. (1992), "The symbolic/subsymbolic interface: hierarchical network organizations for reasoning," in [Sun, Bookman, and Shekhar, 1992] at pp. 17-22.

Lacher, R. C., Hruska, S. I., and Kuncicky, D. C. (1992), "Backpropagation learning in expert networks," *IEEE Trans. on Neural Networks*, vol 3(1), pp 62-72.

LAM (1989), "Computer Software for Safety Glazing" copyright by E. I. du Pont de Nemours & Co., Inc

Lange, T. (1991), "Hybrid connectionist models: temporary bridges over the gap between the symbolic and the subsymbolic," *Technical Report UCLA-AI-91-04*, Computer Science Department, University of California, Los Angeles.

Lange, T. (1992), "Issues in controlling activation and inferencing for natural language understanding in structured connectionist networks," in [Sun, Bookman, and Shekhar, 1992] at pp. 31-38.

Lawrence, J. (1991), "Data Preparation for Neural Networks," AI Expert, vol 6, (November), pp. 34-41.

Lee, J. S. J., and Nguyen, D. D. (1991), "Adaptive algorithms and multilevel processing control in intelligent systems," in [Soucek, 1991] at pp. 407-434.

Lehnert, W. (1991), "Symbolic/subsymbolic sentence analysis: exploiting the best of two worlds," in [Barnden, 1991] at pp 135-164.

Lind, N. C. (1988), "A Design Formula for Glass in Buildings." *Canadian Journal of Civil Engineering*, vol 15, pp.1028-1034.

Lippmann, R. P. (1987), "An Introduction to Computing with Neural Nets." *IEEE ASSP Magazine*, April, pp.4-22.

Lorczak, P. R., Caglayan, A. K. and Eckhardt, D. E. (1989), "A Theoretical Investigation of Generalized Voters for Redundant Systems," FTCS 19. Chicago, IL.

LSGA (1985), *LSGA Standards Manual*, Laminators Safety Glass Association, Topeka, KA.

Machado, R. J., and Rocha, A. F. (1992), "A hybrid architecture for fuzzy connectionist expert systems," in [Kandel and Langholz, 1992] at pp. 136-152.

Maren, A., Harston, C. T., and Pap, R. (eds.) (1990), *Handbook of Neural Computing Applications*, Academic Press, San Diego.

Marr, D., (1982), *Vision*, W.H. Freeman Press, New York.

Marr, D. and Hildreth, E. (1980), "Theory of edge detection," Proc. of Royal Soc. London, vol. B-207, pp. 187-217.

Mazzu, J. M., Caglayan, A. K., and Allen, S. M. (1992), "A hybrid neural network knowledge base for intelligent structural monitoring," Wright-Patterson AFB Report 89171, July.

Mazzu, J. M., Snorrason, M. S., and Caglayan, A. K. (1992), "A hybrid neural network expert system approach to remote sensing," Charles River Analytics Reort 91391, August.

Mazzu, J. M., Caglayan, A. K., and Jonas, M. R. (1991), "A hybrid neural network/expert system environment," Charles River Analytics NASA Contract No. NAS9-18479, August.

McCluney, R. (1993), "Choosing the best window for hot climates," Proc. Innovative Housing '93 Conference, Vancouver, June, FSEC Report CR-577-93.

Medsker, L. R. (ed.), (1991), Special Issue of *Expert Systems with Applications: An International Journal*, vol 2, no. 1.

Medsker, L. (1993), "Models and Techniques for Developing Hybrid Neural Network and Expert Systems," *Journal of Computer & Software Engineering*, to be published.

Medsker, L., and Bailey, D. (1992), "Models and guidelines for integrating expert systems and neural networks," in [Kandel and Langholz, 1992] at pp. 154-171.

Medsker, L. and Liebowitz, J. (1994), *Design and Development of Expert*

Systems and Neural Networks, Macmillan Publishing Company, New York.

Medsker, L. and Turban, E. (1992), "Neural Computing and Artificial Intelligence," in [Turban, 1992] at pp. 621-663.

Medsker, L. and Turban, E. (1994), "Integrating expert systems and neural computing for decision support," *Expert Systems with Applications*, to be published.

Medsker, L., Turban, E., and Trippi, R., (1993), "Neural network fundmentals for financial analysts," *The Journal of Investing*, vol 2, no. 1, pp. 59-68.

Minor, J. E. (1986), "Tests Establish Laminated Glass Strengths." Glass Digest, vol 65, no. 8, pp.74-76.

Minor, J. E., and Behr, R. A. (1993), "Architectural glazing systems in hurricanes," *7th U.S. National Conference on Wind Engineering*, Los Angeles, June, in press.

Minsky, M., and Papert, S. A. (1969), *Perceptrons*, MIT Press, Cambridge, MA.

Minsky, M. (1991), "Logical versus analogical or symbolic versus connectionist or neat versus scruffy," *AI Magazine*, vol 12, no. 2, pp. 35-51.

Nelson, M. M. and Illingsworth, W. T. (1991), *A Practical Guide to Neural Nets*, Addison-Wesley Publishing Company, Reading, MA.

Norville, H. S. (1993), "Technics focus: wind does not break window glass," *Progressive Architecture*, April, pp. 148-152.

Oxman, S. W., Dockery J. T. (1990), "The Synthetic Expert," *PC AI*, vol 4, no. 1, pp.50-51.

Pinkas, G. (1992), "Representing unrestricted first-order logic formulas in connectionist networks," in [Sun, Bookman, and Shekhar, 1992] at pp. 23-30.

Posey, C., Kandel, A., and Langholz, G. (1992), "Fuzzy hybrid systems," in [Kandel and Langholz, 1992] at pp. 174-197.

Rajapakse, J., and Acharya, R. (1990), "MultiSensor data fusion within hierarchical neural networks," *Proc. of the IJCNN*, San Diego, vol. II, (June), pp. 17-22.

Rasmussen, J. (1985), "The Role of Hierarchical Knowledge Representation in Decision Making and System Management." *IEEE Transactions on Systems, Man, and Cybernetics*, vol 15, no 2.

Reid, K., and Zeichick, A. (1992), "Neural Netwrok Resource Guide," *AI Expert*, vol 7, no. 6, pp. 50-56.

Resnick, P. L. and Minor, J. E. (1986), "Failure Strength of Laminated Glass Units." Glass Research and TestingLaboratory, Texas Tech University, Lubbock. (NTIS Accession No. PB87-118873/AS).

Romaniuk, S. G., and Hall, L. O. (1991), "Injecting symbol processing into a connectionist model," in [Soucek, 1991] at pp. 383-405.

Rosenblatt, F. (1962), *Principles of Neurodynamics*, Spartan, New York.

Roth, M. W. (1988), "Neural-Network Technology and Its Applications," *Johns Hopkins APL Technical Digest*, vol 9, no. 3.

Roth, M. W. (1990), "Survey of neural network technology for automatic target recognition," *IEEE Trans. on Neural Networks*, vol 1, No 1, pp.

Ruch, J. C. (1984), Chapter 2 in *Psychology: The Personal Science*, Wadsworth Publishing Company, Belmont, CA.

Rumelhart, D. E. *et al.* (1986), "Learning internal representations by error propagation," *Parallel Distributed Processing*, vol. 1, McClelland, J.L. and Rumelhart, D.E. (eds.), MIT Press, Cambridge, pp. XX-YY.

Samad, T. (1992), "Hybrid distributed/local connectionist architectures," in [Kandel and Langholz, 1992] at pp 200-219.

Shafer, G. and Logan, R. (1987), "Implementing Dempster's rule for hierarchical evidence," *Artificial Intelligence*, vol 33, pp. 271-298.

Sharkey, N. E. (1992), "The expanding role of superpositional representations in hybrid systems," in [Sun, Bookman, and Shekhar, 1992] at pp. 54-57.

Sherald, M. A. (1989), "Neural Networks versus Expert Systems," *PC AI*, vol 3, no. 4, pp. 10-15.

Soucek, B., and the IRIS Group (eds) (1991), *Neural and Intelligent Systems Integration*, John Wiley and Sons, New York.

Sowers, R. M. 1981. "Laminated Materials, Glass." *Encyclopedia of Chemical Technology, Third Ed.*, vol 13, John Wiley and Sons, New York, pp. 978-993.

Stickel, M. E. (1985), "Automated deduction by theory resolution," Journal of Automated Reasoning, vol 1(4), pp 333-355.

Stolke, A. and Wu, D. (1992), "Tree matching with recursive distributed representations," in [Sun, Bookman, and Shekhar, 1992] at pp. 58-65.

Sumida, R. A. (1992), "PDS networks: integrating localist and PDP approaches for natural language processing," in [Sun, Bookman, and Shekhar, 1992] at pp. 46-53.

Sun, R. (1991), "Integrating rules and connectionism for robust reasoning: a connectionist architecture with dual representation," Ph.D. dissertation, Brandeis University.

Sun, R. (1992), "Connectionist models of rule-based reasoning," in [Sun, Bookman, and Shekhar, 1992] at pp. 91-97.

Sun, R. and Bookman, L. (1993), "How do symbols and netwroks fit together," AI Magazine, vol 14, no. 2, pp. 20-23.

Sun, R., Bookman, L. A., and Shekhar, S. (eds.) (1992), *Integrating Neural and Symbolic Processes: The Cognitive Dimension*, Working Notes from the AAAI-92 Workshop Program, San Jose, CA, July 12-16, 1992.

Sun, R., and Waltz, D. (1991), "A neurally inspired massively parallel model of rule-based reasoning," in [Soucek, 1991] at pp. 341-381.

Towell, G., and Shavlik, J. W. (1992), "Interpretation of artificial neural networks: mapping knowledge-based neural networks into rules," in Advances in Neural Information Processing Systems, vol 4, Morgan Kaufmann, pp. 977-984.

TR-440 (1983), "Guidelines for the testing and evaluation of products for cyclone-prone areas," Experimental Building Station, Dept. of Housing and Construction, Chatswood NSW, Australia, March reprint.

Turban, E. (1992), Expert Systems and Applied Artificial Intelligence. New York: Macmillan Publishing Company.

Vallabhan, C. V. G., Das, M., Magdi, M., and Bailey, J. R. (1991), "Experimental verification of the theoretical solution of laminated glass units," *Proc. ASCE Specialty Conference, Advanced Composites in Civil Engineering Structures*, Las Vegas, January.

Vild, D. J. (1993), "Glass and wind loads," *Glass Digest*, July 15, pp. 36-37.

Villa, M. F., and Reilly, K. D. (1992), "Hierarchical structures in hybrid systems," in [Kandel and Langholz, 1992] at pp. 222-254.

Wasserman, P. (1989), *Neural Computing: Theory and Practice*, New York: Van Nostrand Reinhold.

Werbos, P. (1989), "Back-Propagation and Neuro-Control: A Review and Prospectus," *Proceedings, Joint Conference on Neural Networks*, Washington, D.C., June.

Werbos, P. (1990), "Neurocontrol and related techniques," in [Maren, Harston, and Pap, 1990] at pp. 345-380.

Wertbos, P. (1993a), "Brain-like intelligence in artificial models: khow can we really get there?" *INNS Above Threshold*, vol 2, No. 2, pp 8-12.

Werbos, P. (1993b), *The Roots of Backpropagation: From Ordered Derivatives to Neural Networks and Political Forecasting*, Wiley, New York.

Wermter, S., and Lehnert, W. G. (1989), "A hybrid symbolic/connectionist model for noun phrase understanding," *Connection Science*, vol 1 (3), pp. 255-272.

Wermter, S. (1992), "SCANning understanding: a hybrid and connectionist architecture," in [Sun, Bookman, and Shekhar, 1992] at pp. 83-90.

Wilson, A., and Hendler, J. (1993), "Linking symbolic and subsymbolic computing," *Technical Research Report TR 93-12*, Institute for Systems Research, University of Maryland.

Yang, Q. (1992), "Rule combining - a neural network approach to design optimization," in [Kandel and Langholz, 1992] at pp. 256-275.

Ye, N., McAvoy, T., Kosanovich, K., and Piovoso, M. (1993), "Plant-wide control using an inferential approach," Technical Report, Department of Chemical Engineering, University of Maryland.

Zadeh, L.A. (1965), "Fuzzy sets," *Information and Control*, vol. 8, pp. 338-353.

Zadeh, L.A. (1986), "A simple view of the Dempster-Shafer theory of evidence and its implication for the rule of combination," *AI Magazine*, pp. 85-90.

Index